Pollution-Pricing

Pollution-Pricing

Industrial Response to Wastewater Charges

James F. Hudson
Elizabeth E. Lake
Urban Systems Research
and Engineering, Inc.
Donald S. Grossman

LexingtonBooks
D.C. Heath and Company
Lexington, Massachusetts
Toronto

Library of Congress Cataloging in Publication Data

Hudson, James F 1950-
Pollution-pricing.

Includes bibliographical references and index.

1. Sewage disposal—Economic aspects—United States. 2. Factory and trade waste—Economic aspects—United States. 3. Environmental protection—Economic aspects—United States. I. Lake, Elizabeth E., joint author. II. Grossman, Donald S., joint author. III. Title.
TD523.H83 363.7'3946 80-8363
ISBN 0-669-04033-9

Published simultaneously in Canada

Printed in the United States of America

International Standard Book Number: 0-669-04033-9

Library of Congress Catalog Card Number: 80-8363

Contents

List of Figures

List of Tables

Preface

Research in the field of pollution economics has suggested that major improvements in environmental programs could be achieved by using economic incentives to supplement or partially replace regulations. Much of the past research, however, has been concerned either with the aggregate response of large numbers of industrial plants to incentives or with the theory surrounding optimal response. The actual responses of individual plants that are facing economic incentives to reduce residuals discharge have not been adequately studied.

The study on which this book is based investigated such responses and analyzed the operation of local charges for pollutants in wastewater. Through in-depth case studies of five charge systems and 101 industrial plants, both charge-system operation and the methods used by plant managers to reduce their charges were analyzed. In each of the charge systems, local wastewater fees were based on the quantity and quality of discharges to the sewers and public treatment facilities. Administrative cost, monitoring accuracy, effectiveness in cost allocation, and acceptance by users were key concerns addressed in the study.

From the perspective of the plant managers at the industrial facilities, responses could be categorized into *administrative* changes that alter only the amounts charged, and *actual* changes in discharge quality or quantity, also leading to reduced charges. The actual changes resulted from improved housekeeping, wastewater treatment, water reuse, and other modifications in inputs, products, and production methods.

In all five systems, the charges were generally accepted and operated successfully; almost all plant managers noticed the charges and attempted to reduce them. The actual reductions in discharges were not large, but neither were the charges. Thus, the basic results of the study are that the charge systems worked from the viewpoint of their operators and users and that the users showed an economic response, changing their behavior to reduce costs. The context—the set of system rules and regulations, other environmental requirements, and other factor prices—affected responses considerably, but the basic result held throughout. The implication, within the limits of this analysis, is that economic incentives are practical in pollution control.

Acknowledgments

This book would not have been possible without the cooperation and support of the systems in the five cities studied. Our main acknowledgment, therefore, must go to:

Pollution Control Department, Atlanta, Georgia

Metropolitan Sanitary District of Greater Chicago

Dallas Water Utility, City of Dallas, Texas

Department of Public Works, City Hall, Salem, Oregon

Public Services Department, South San Francisco, California

The staff of all five systems were extremely cooperative, and their support made the project possible. Similarly, the willingness of plant managers and officials in the 101 industry case studies was critical. Although we are unable to disclose their names because of restrictions on confidentiality of the data, we would like to give them our sincere thanks for the information they provided on the workings of the charge systems.

Many helpful comments throughout the project were received from the Council on Environmental Quality, represented by Gordon Brady and Blair Bower, and by our consultants, James Seagraves, Clark Binkley, and Harris Gold. The staff members of Urban Systems Research and Engineering, Inc., and Resource Analysis, Inc., who participated extensively in this project, are Patricia Deese, David Gaboury, Susan Farrell, Malcolm Persen, Ellen Kane, Jo Bachelder, and Carl Pechman. They all deserve credit for their successful work on the project. Their comments and support have been of great assistance, though we must, of course, take full responsibility for the final interpretations and conclusions and the conduct of the work as a whole.

This work was supported by the President's Council on Environmental Quality under contract EQ8AC029. Portions of the final report of that contract are reprinted with permission.

List of Abbreviations

BAT	best available technology
BOD	biochemical oxygen demand
BPT	best practical technology
ccf	hundred cubic feet
CEQ	President's Council on Environmental Quality
CIP	cleaned in place
COD	chemical oxygen demand
DEQ	Department of Environmental Quality
EPD	Environmental Protection Division
FOG	fats, oils, and grease
mgd	million gallons per day
mg/l	milligrams per liter
NPDES	National Pollutant Discharge Elimination System
NSPS	new source performance standards
O&M	operation and maintenance
ppm	parts per million
SS	suspended solids
TSP	total suspended particulates
TSS	total suspended solids

Part I
Major Findings

1 Introduction

Historically, local governments responsible for the provision of sewerage services financed many of their capital, operations, and maintenance costs through property taxes. This is in strong contrast to financing of water supply, which has almost always been on a fee-for-service basis. More recently, these governments have shifted to sewer charges in order to augment their revenues and to ensure that waste treatment costs were fairly allocated among the users of the sewer systems. In 1945, fewer than 20 percent of large (over 50,000 population) U.S. cities levied some form of sewer charges. By 1970, this proportion was almost 90 percent. By 1971, nearly two-fifths of local expenditures on sewerage were derived from charges, and the user-charge provisions of the Clean Water Act have raised this proportion.

In addition to raising revenues and attempting to ensure that the distribution of costs between users is fair, sewer charges represent a price for services and, to the user, a cost that must be paid for each unit of discharge, therefore becoming an incentive to reduce pollution. Economists have long claimed that water-pollution problems result from a failure in the market system. Because in the past the residuals of production and consumption activities could be freely dumped into water bodies, undesirably high levels of water pollution have resulted. There are parallels for air pollution and solid-waste management as well. In response to these situations, some economists suggest that pricing the use of assimilative capacity in the environment would correct the market failure and consequently would lead to improved environmental quality (at least if prices were set properly).

For this reason, the practical experience of discharges to operators of systems in communities that have imposed sewer charges is relevant to the evaluation of economic incentives in environmental quality management. This is particularly true where the charges have been based both on flow and on the quantities of pollutants. Current federal policy initiatives seek to supplement or replace regulatory approaches to environmental-quality management with economic incentives, particularly in the areas of air quality, water quality, and solid-waste management. To support these initiatives, the experience with existing programs where charges such as the sewer charges are used needs to be reviewed and analyzed.

There is no shortage of literature describing the theoretical use of economic incentives for pollution control. There is also a body of literature describing empirical studies of waste reduction associated with sewer charges.[1] Most of these studies, however, have analyzed the response of groups of users in a particular industry without looking closely at the decisions made at individual plants. Most also fail to consider the institutional arrangements of the agencies responsible for the imposition and administration of the charge system. This book seeks to correct these shortcomings through an analysis of charge systems in five cities: Atlanta, Georgia; Chicago, Illinois; Dallas, Texas; Salem, Oregon; and South San Francisco, California.

The results obtained should provide some practical insights into the effectiveness of economic incentives for pollution control and into the problems such incentives can represent.

In each city, representatives from the agencies responsible for the charge program were interviewed to learn about the specific objectives of their programs; the characteristics of both the wastewater system and the charges; the administration of their programs, including monitoring, billing, and enforcement practices; and their success in achieving the stated objectives. In addition, this survey looked at the context for the charges—the numerous regulations on prohibited discharges, required treatment, solid-waste disposal, and the like—that constrain the potential plant responses to the charges. The results of this effort are summarized in chapter 2 and described in more detail in part II.

In addition, approximately one hundred industrial users in these five systems were selected for analysis of their responses to the charges. The responses covered a wide range, from simple administrative attempts to reduce billings without any real physical changes through such things as shifts in how wastewater is measured. Some plants also responded by reducing waste discharges through pretreatment in-plant modification, housekeeping improvements, or changes in the line of products or the inputs used. Responses were affected by other factors, such as increased energy prices, and the presence of other, sometimes conflicting, government regulations. Plant managers, engineers, and financial officers were interviewed at each of the plants to collect information about their operations and the responses they selected. The results obtained through the plant surveys are presented in chapter 3 and sample cases presented in part III.

The experiences of both the agencies responsible for the sewer-charge programs and the discharges on which the charges are imposed offer some lessons about the utility and the limitations of these charges and therefore of other incentive systems. These policy implications are discussed in chapter 4.

Note

1. See appendix for bibliography of other literature in the field.

2 Sewer Charges for Industrial Wastes: The Experience of Five Cities

This chapter describes how the five cities studied implement their industrial sewer-charge systems. Industrial sewer charges often have two components, one related to the volume of flow and the other to the strength of the wastes included. The two components together are typically designed to recover the costs of treating industrial wastewater. (As a further complication, many systems separate the fees into a basic *charge,* which pays for the flow-related portion and a nominal level of pollutants, and a *surcharge,* which pays for high-strength wastes.) Industrial waste flows respond to both components of the charge.

Table 2-1 presents some summary data about the systems studied. All have been in operation long enough for reasonable conclusions to be drawn from their experience. The charge system in South San Francisco is the oldest of the five, having been in existence since 1951; the newest system was put into operation in 1970. Collectively, these systems range from the quite small one in South San Francisco to one of the largest sewer agencies in the world, the Metropolitan Sanitary District of Greater Chicago. Four are departments of municipal governments and one (Chicago) is a special sewer district.

In all cases, the objectives of the sewer surcharge programs are the same: the provision of adequate revenues to finance fixed and variable system costs and equitable distribution of the system costs among users. The charges were not designed as incentives for flow and concentration reductions.

Table 2-1
Summary Data for Wastewater Systems

City	*Year of Sewer Charge Adoption*	*Capacity of System (mgd)*[a]	*Administrative Unit*
Atlanta, GA	1967	130	Municipal
Chicago, IL	1970	1,500	Special district
Dallas, TX	1959	138	Municipal
Salem, OR	1965	35	Municipal
South San Francisco, CA	1951	9[a]	Municipal/regional

[a] Capacity of sewage treatment plant(s) as of 1979.

Despite the fact that the potential incentive properties of a sewer charge were generally ignored in the cities examined, economic calculations (least-cost solutions) seemed to govern the response of individual plants to the charges levied. Because the level of the sewer charges is usually low compared to the cost of waste reduction or self-treatment, these economic calculations generally indicated that only modest reductions in waste flow or strength could be justified. Consequently, total flows into municipally owned treatment works were not perceptibly affected by the imposition of sewer charges, and the impacts on treatment plant operations were minor. In the South San Francisco case study, the drought of 1976–1977 led to significant reductions in the volume of wastewater entering the plant, but operating costs were not measurably altered. Hence, one possible impact of sewer charges—changes in the costs of municipal waste treatment due to changes in the levels of waste flow—was not observed in this research. In any case, this would probably be a long-term response arising from reduced need for new capacity, not a reduction in annual operation and maintenance (O&M) costs. Furthermore, in no case were either business closings or shifts to self-treatment and discharges with National Pollutant Discharge Elimination System (NPDES) permits observed.[1] Chapter 3 discusses industrial response in greater detail.

The remainder of this chapter treats specific aspects of industrial sewer charge systems. The first section describes the design of the charge systems, focusing on the costs and methods of monitoring discharges. The following section discusses the financial aspects of the charge systems.

Design of the Charge Systems

The five cities studied had common objectives for their sewer charge systems: the recovery of the local costs of wastewater management and an equitable distribution of these costs among users. However, the details of implementing these objectives differ significantly among cities. The differences include the cost components considered, the treatment of increased future costs of replacing existing systems, and the allocation of costs between waste components (flow, BOD or COD, and TSS). The concern here is with the design of sewer charge systems, including the costs considered, the form of the charge, and the number of plants covered in each city.

In all of the cities except Atlanta and South San Francisco the sewer charge system was designed to cover the costs associated with all phases of sewage collection, sewage treatment, and sludge disposal. In Atlanta, for example, capital and operating costs for interceptors were financed from general revenues by a separate Bureau of the Department of Environment and Streets and consequently are excluded from the sewer charge cost basis.

Because interceptors generally comprise well over half of the total capital costs of a wastewater management system, this omission seriously understates the true cost of sewer services to those who pay sewer charges.

In setting the charge levels, it is necessary to allocate the costs of various processes for collection, treatment, and sludge disposal to the various waste components. This cost allocation involves some use of engineering judgment because many wastewater treatment processes affect several waste components. Although cost allocation guidelines are available, there is a degree of arbitrariness in any setting of rates.[2] In three of the cities—Dallas, Salem, and South San Francisco—total annual costs for the following year (both O&M and debt service) are projected. Then the costs associated with collecting and treating a unit of volume, BOD or COD, and TSS are determined by considering each component of the process (collection, transmission, grit removal, clarification, aeration, and so forth) and allocating some fraction of the cost of each component to each portion of the waste stream. Table 2–2 shows a typical allocation, using Dallas as an example. In Atlanta, actual costs from the previous year are used in setting the level of the unit charge for each component in the present year. To allocate costs between waste components, Atlanta's agency assumes the cost of treating a unit of BOD equals that of treating a unit of TSS. In Chicago, the costs allocated to suspended solids are estimated from actual system costs, the

Table 2–2
Allocation of Functional Costs to Service Components in Dallas

	Allocation of Costs (Inside City Only)			
Function	*Volume*	*BOD*	*SS*	*Customer Service*
Collection treatment:	100	0	0	0
Pumping	100	0	0	0
Biological treatment	0	100	0	0
Grit removal	8	0	92	0
Clarification	24	37	38	0
Chlorination	100	0	0	0
Sludge disposal:				
Pumping, chemicals, and digestion	0	0	100	0
Lagoons	0	17	83	0
Customer service	0	0	0	100
Engineering and other administration	72	8	13	7
Percent of total inside city costs	74.2	9.4	9.6	6.7

Source: Turner, Collie and Braden, Inc., *Dallas Utilities Sewer Rate Study*, Table E–2. Dallas, Texas, 1977.

costs allocated to BOD are estimated from design costs, and the costs allocated to volume are estimated from a combination of actual and design costs. In short, cost allocation is not a uniformly defined or practiced procedure.

Dallas is the only city of the five that includes an amount designed to reflect the higher future costs of replacement as a component of its sewer-charge system. In Dallas and Atlanta, the terms of the revenue bonds used to finance the construction of the system require that revenues exceed debt-service requirements by some amount, ranging from 33 percent to 50 percent. This provision could automatically provide some measure of protection against increased future costs if those funds were retained in a sinking fund earmarked for sewerage. In Atlanta, though, revenue excesses are returned to the city's general fund.

Tables 2-3 and 2-4 show the cost formulas and cost factors for the five cities. In brief, Dallas and Atlanta have charge and surcharge systems as described earlier, with the volume charge assumed to include a normal level of wastes and a surcharge applied to wastes of excessive strength. South San Francisco has no normal level and charges directly on pounds of BOD and TSS. Chicago raises most of its revenues from ad valorem taxes paid to the sewer district; however, plants whose waste-handling cost to the district is estimated to be higher than their tax payments pay that higher cost instead.

Table 2-3
Structure of the Charge Formulas

$$\$/\text{ccf} = V + A\,(C - C_0)\,\text{TSS}\;A\,(B - B_0)\,\text{BOD}$$

where:
- V = volume-related charge ($/ccf)
- TSS = TSS-related charge ($/lb)
- BOD = BOD-related charge ($/lb)[a]
- C = actual TSS concentration (mg/l)
- C_0 = normal TSS concentration (mg/l)[b]
- B = actual BOD concentration (mg/1)
- B_0 = normal BOD concentration (mg/l)[b]
- A = conversion factor from mg/1 to lb/ccf = 0.006238

In Atlanta, the formula is:

$$\$/\text{ccf} = \$.88 + \$.105 \times .006238 \times [(C - 250) + (B - 250)]$$

[a] COD in South San Francisco.

[b] 250 mg/1 in Atlanta and Dallas; 0 elsewhere.

Table 2-4
Cost Factors for the Charge Formulas

City	*BOD*	*TSS*	*FLOW*[a]
Salem	\$.0181/lb (+\$.00161/lb peak)[d]	\$.0210/lb (+\$.00038/lb peak)	\$.059/ccf[b] (+.00167 peak)
Dallas	\$.025/lb	\$.025/lb	\$.28/ccf[c]
South San Francisco	\$.034/lb(COD)	\$.074/lb	\$.31/ccf
Chicago	\$.041/lb	\$.056/lb	\$.031/ccf
Atlanta	\$.105/lb	\$.105/lb	\$.88/ccf

[a] In ccf based on estimated discharges.

[b] An additional charge is imposed on seasonal flows.

[c] A rate of \$.30/ccf is charged for metered discharges of wastewater, as opposed to estimates based on metered inflow to the plant.

[d] This rate is applied to the discharge in the peak month for the plant. Each month, the plant pays \$.0181/lb for BOD discharges then, plus \$.0016/lb for pounds discharged in the peak month of BOD discharge in the previous year. The fomula is similar for TSS and flow, although there can be three different peak months—one for each pollutant.

In Salem, there are two types of charges: a base charge, similar to that in South San Francisco, and a peaking or "demand" charge. The latter essentially charges plants according to the portion of system capacity they use in peak periods.

An example of the use of the formula and charges would be useful here. Assume we have a plant in Atlanta that discharges waste with a TSS level of 340 mg/l and a BOD level of 460 mg/l. Then the charges would be:

$$(340 - 250) + (460 - 250) \times 0.006238 = 1.87 \text{ pounds/ccf} \quad (2.1)$$

$$1.87 \text{ pounds/ccf} \times \$.105/\text{pound} = \$.196/\text{ccf} \quad (2.2)$$

$$\$.196/\text{ccf} + \$.88/\text{ccf} = \$1.08/\text{ccf (final charge rate)} \quad (2.3)$$

The first step was to compute the weight of BOD and TSS subject to the charge (above the normal level). This was converted to a volume-related fee in the second step, which was added to the base fee in the third. Thus, this plant would pay \$1.08/ccf for wastewater service in Atlanta.

In some cities, the cost factors have been relatively stable since the program's initial implementation. In Dallas, there were no rate increases between 1959 and 1976, although there have been two rate increases since then. Salem had no rate increases between 1965 and 1976. On the other hand, Atlanta, Chicago, and South San Francisco have seen frequent rate increases. Atlanta had six increases between 1967 and 1978, while Chicago and South San Francisco have had almost annual increases since 1968, although the latter had only two increases during the previous twelve years of the program.

In all cases, only industrial users are subject to the charge and surcharge levels described here. However, in Dallas and Atlanta, any plant with effluents above the defined normal concentrations are included in this category as "industrial" by definition. In Dallas, certain types of businesses pay a "class-rate" surcharge by which, for example, a set of restaurants is sampled for BOD and TSS concentrations and the resulting surcharge rate is applied to all restaurants. Variances are possible, but the goal of this class surcharge was to improve cost allocation without the expense of monitoring several thousand discharges. Of those that are monitored, there is a wide variation in charges paid (see table 2-5).

Salem has a similar system: all plants discharging more than 200 pounds of BOD and TSS per day must be monitored and billed under the system. Those with discharges of 50 to 200 pounds of BOD and TSS per day may or may not be monitored, at the discretion of the director. The unmonitored plants in this category are billed according to average concentrations, as in Dallas. In South San Francisco, the industrial categories to be monitored and the charges are defined by a city ordinance. Institutions, offices, and restaurants pay differing volume charges, reflecting average concentrations from published tables. In Chicago, the direct charge is applied only to manufacturers; all others pay only the ad valorem assessment.

Typically only a small proportion of dischargers is affected by the strength-related charges. Table 2-6 presents some data on the number of plants covered in each of the study cities. Although the number of dischargers included in these programs is small (ranging from 24 in South San Francisco to 350 in Chicago), the total amount of waste load and flow collectively presented by these dischargers is substantial.

In general, only very few dischargers are specifically exempted from the provisions of the user charge/surcharge programs. In Dallas, even the city government is included in the program. In Atlanta, federal installations fall under the sewer-charge program. At present, a large restaurant and catering operation in Chicago falls outside the surcharge system, although their waste flows are large enough to qualify them for inclusion. This situation will be rectified in a forthcoming revision of the program, which will extend it to include nonmanufacturing plants.

Finally, the sewer-charge systems analyzed in this study are complemented by regulation of wastes other than BOD and TSS. Most regulate

Table 2-5
Percentage Distribution of Strength Charges for Monitored Plants in Dallas

Strength Charge ($/mo)	*Number of Plants*	*Percentage of Plants*
0	19	10.2
1-250	81	43.5
251-500	15	8.1
501-1000	19	10.2
1001-1500	30	16.1
1501-2000	16	8.6
2000 +	6	3.2
	186	
Mean Charge		$611.69/month
Median Charge		$458.49/month

fats, oils, and grease (FOG), pH, temperature, explosives, radioactive materials, noxious substances, garbage and solids that may obstruct sewers and treatment facilities, metals, toxics, sulfides, and other materials. This extensive coverage and similarity undoubtedly stems from the common purpose of municipal sewage-treatment works. In each of the cities, monitoring activities cover the prohibited and regulated wastes as well as the BOD, TSS, and COD concentrations that form the basis of the charges.

Administration

Administration of a sewer-charge system involves monitoring the volume and strength of wastes in order to set the amounts of the charges to be paid and enforce the provisions of the charge system. This section discusses the methods, costs, benefits, and performance of these activities in the study cities.

In four of the five cities, an agency of the municipal government is responsible for monitoring quantity and quality of discharges for the purpose of setting the sewer-charge level. Chicago alone supports a system of industrial self-monitoring, with spot checks by the district's own monitoring team. Large dischargers are visited about once a year under this program. Atlanta is considering replacing its program of municipal monitoring by a self-monitoring system.

Table 2-6
Coverage of Pollutant-Strength Charges

City	Total Number of Customers	Eligible for Charge		Actually Charged		Percentage of Total BOD Load	Percentage of Total TSS Load	Percentage of Total Flow
		Number	Percentage	Number	Percentage			
Atlanta	100,000	200	0.20	66	33.0	NA	NA	15
Chicago	2,000,000	2000	0.10	350	17.5	19	8	6
Dallas	229,000	185	0.08	119	64.3	50	25	10
Salem	50,000	50	0.10	18	36.0	39 (peak = 42)	18 (peak = 24)	9 (peak = 30)
South San Francisco	20,000	24	0.12	24	100.0	47[a]	34	13

[a] Percentage of COD load.

The details of the monitoring programs differ considerably across the five cities. The Chicago program calls for hourly sampling for a consecutive seven-day period once per year. In Dallas and Atlanta, two samples are to be taken yearly, but Dallas takes daily composites for a week, while the sample in Atlanta may be a simple grab.[3] South San Francisco takes four daily-composite samples annually. The sampling frequency in Salem is not specified by ordinance, but the agency takes daily-composite samples at a reasonably high frequency. None of the systems requires continuous sampling and analysis, although Salem requires that the sample be taken by the firm and be ready for analysis if the city chooses.

Except in Salem, wastewater flow is estimated from water flows, unless the discharger chooses to install a wastewater flow meter or can justify an estimate of the amount of water consumed in the plant or shipped out with the product. In Salem, wastewater flow is measured continuously as part of the monitoring program.

A number of different procedures are followed in case of disputes concerning values determined in the sampling program. Atlanta will take a second sample at the request of the discharger (who pays the cost) and will accept the lower of the two samples. This procedure can be expected to bias the estimated sewer charge downward. Dallas will also take a second sample, paid for by the discharger, but will use the new value. Discrepancies in Chicago's sample, which are split with the industry, result in negotiations concerning the level of the charge. In Dallas, a third sample may be taken after disagreement and analyzed by an independent lab; whichever estimate (the city's or the firm's) is closer to the independent lab's estimate will be used to set the level of the charge, although this provision has not yet been used.

In Atlanta, Salem, South San Francisco, and Chicago, no advance warning is given prior to the visit by the monitoring team. In Dallas, one week's notice is generally given to ensure that a suitable monitoring site is available and that the firm will be in operation. Unannounced spot checks are made, however, particularly where there may be discharges of prohibited wastes. The effect of monitoring on industry behavior is discussed in chapter 3.

Table 2-7 summarizes the administrative costs of the industrial sewer-charge systems in the study cities. The lowest cost per eligible plant was reported in Chicago, which uses a self-monitoring system and thus puts the cost burden on the plants. Since only a small portion of the eligible plants in Chicago actually pay a charge, the administrative cost per charged plant is high, ranging from $2,100 to $3,400 per plant charged. In the cities where government monitors industrial discharges, the cost per eligible discharge ranges from $500 to $1,300. Salem reports the lowest cost per charged plant, but that figure does not include the costs to the private sector of providing an approved sampling site and facilities to keep the sample refrigerated. Those capital costs were estimated at $6,000 for each discharger. The costs of administering the industrial charge program (including monitoring) ranged from 7 percent to 21 percent of the revenues generated by the program.

The monitoring programs have had beneficial side effects beyond implementing the provisions of the industrial sewer-surcharge programs. For example, in Dallas the monitoring crew was able to trace a waste discharge that was diminishing the removal efficiency of the treatment facility back to the plant responsible for the discharge. Prior to establishing the industrial program, the city would not have had the capability to do so. Similar problems have been discovered in routine monitoring. Data collected under both limited-substances ordinances and the industrial charge programs have generally given the treatment-plant operators a better idea of where to locate dischargers whose wastes adversely affect the treatment processes.

Enforcement has not been a significant problem in any of the cities; litigation and service shutoff are virtually nonexistent. However, establishing the programs has not always been easy. In Chicago, for example, the legality of the charges was initially appealed to the Illinois Supreme Court.

The lack of enforcement problems may have resulted from the efforts made by the communities to enlist the support of the business community for the surcharge programs. For example, representatives of the Dallas utility attend local manufacturers' trade-association meetings to explain the provisions of the program. Public meetings are held on all rate increases, and the staff has worked with individual plants to explain the provisions of the surcharge program. The South San Francisco utility keeps close contact with the Peninsula Manufacturing Association to discuss the surcharge program. The other cities have not yet made such extensive efforts to obtain and keep the support of industry, although rate increases are generally subject to public meetings.

Financing of Operations

The beginning of this chapter indicated that financial self-sufficiency is the stated objective for the sewer (and water) services in all five study cities. In several instances, this is a requirement for the bonds issued to finance capital costs. However, it is difficult to assess the extent to which this self-sufficiency is met, because in three cities—Atlanta, Dallas, and Salem—water and sewer costs are in combined accounts, particularly the capital costs of

Table 2-7
Administrative Costs of Industrial Charge Programs

	Administrative Cost			*Total Pollutant Revenue Charge (Millions of Dollars)*	*Administrative Costs as a Percentage of Charge Revenues*
City	*Thousands of Dollars per Year*	*Dollars per Eligible Plant*	*Dollars per Plant Charged*		
Atlanta	100	500	1500	0.47	21
Chicago	750–1200	375–600	2100–3400	5.70	13–21
Dallas	245	1300	2100	1.30	19
Salem	50	1000	2800	0.76	7
San Francisco	NA	NA	NA	0.53	NA

Note: Includes costs of monitoring, record-keeping, billing, and analysis of samples.

water and sewer bonds. This section first examines the data on self-sufficiency and then considers the impact of industrial charge programs on the revenues of the agencies studied.

Income statements for the agencies have been constructed, as presented in table 2–8. Chicago earns net revenue from its sewer operations. In the case of Chicago, most of the agency's operations are financed out of a special property-tax levy explicitly dedicated to the sewer agency. For any individual discharger, the net charge equals the gross sewer charge less the amount of property tax paid to the sewer agency. In Dallas, Atlanta, and Salem, the combined water and sewer utilities earn net revenue, with debt-service costs included; these excess revenues may, in theory, be accrued for future capital costs. A second concern is whether there are cross-subsidies between water and sewer. Because debt-service costs are not reported independently for the waste and sewer utilities, it is not possible to assess the adequacy of sewer revenues to cover total sewer-related costs. However, further analysis suggests that the sewer agencies alone generate net revenues for the city and to some extent may subsidize the water operations in Atlanta, Dallas, and Salem.

The industrial charge amounts to a small portion of total sewer agency revenues in all of the study cities. This can be seen in table 2–9, which shows that the industry contributed from 10 percent to 26 percent of total sewer revenues. However, this is generally more than the total allocated cost. As well as can be estimated, the plants are paying their own way, which was the basic goal for the systems.

Table 2-8
Sewer/Water Revenues and Costs
(thousands of dollars)

Atlanta, 1977 (Water and Sewer)		*Chicago, 1977 (Sewer Only)*		*Dallas, 1978 (Water and Sewer)*		*Salem, 1976–1977 (Water and Sewer)*		*South San Francisco, 1979 (Sewer Only)*	
Revenues		*Revenues*		*Revenues*		*Revenues*		*Revenues*	
Water	25,909	Property tax	$116,565	Water sales	$35,599	Water	2,444	Residential	713
Sewer contracts	2,833	Grants:		Sewer charges	22,015	Sewer	3,952	Institutional	38
Sewer charges		Federal	108,197	Sewer contracts	717	Other	649	Businesses not	
Surcharge	465	State	5,092	Sewer surcharges	1,225	Total	7,045	monitored	327
Other	488	Industrial surcharge	5,659	Sewer penalties	299			Monitored:	
Total	41,554	Sale of fertilizer	971	Other	7,274			Industrial (volume)	106
		Land rentals	794	Total	67,129			Industrial (surcharge)	420
		Interest on						Total	1,604
		investments	6,725						
		Miscellaneous	4,534						
		Sale of bonds	80,000						
		Total	328,537						
Expenditures		*Expenditures*		*Expenditures*		*Expenditures*		*Expenditures*	
Water operations	13,295	O&M	62,134	Operating:		Administration	1,290	O&M	1,340
Sewer operations	7,661	R&D	5,720	Personnel	18,728	Water operations	1,059	Debt service	231
Debt service	11,034	Administration	14,385	Debt amortization	7,308	Sewer operations	1,470	ICR	33
Total	31,990	Pension	6,765	Other	14,490	Debt service	1,463	Total	1,604
		Bond redemption	20,750	Other:		Water construction	145		
Excess[a]	9,564	Interest on bonds	18,053	Interest on bonds	6,273	Sewer construction	221		
		Capital		Interest on notes	535	Total	5,648		
		expenditures	176,499	Loss on property	34				
		Total	304,306	Total	47,368	Excess	1,397		
		Excess	24,231	Excess	19,761				

[a] Net income; generally applied to future capital funds.

Table 2-9
Industry Contribution to Revenues and Costs

	Atlanta	*Chicago*	*Dallas*	*Salem*	*South San Francisco*
Direct pollutant charge to industry (percentage of all system revenues)	2.5	5.4	5.1	18.0	26.2
Percentage of revenues paid by these plants through other charges[a]	11.3	1.3	9.1	0	0
Total	13.8	6.7	14.2	18.0	26.2
Total industry costs (derived from table 2-5)	15[b]	7.5[c]	14.6[d]	7.8[c]	18.5[c]

[a] Base (<250 mg/1) charges in Atlanta and Dallas; property tax in Chicago.

[b] Volume data only; since industrial strength is higher, this probably underestimates costs.

[c] System costs assumed to be 80-percent flow-related, 10 percent BOD, 10 percent TSS; for Salem, does not include peaking.

[d] Actual costs from Dallas.

Notes

1. A few plants did obtain NPDES permits for the discharge of cooling water. In one instance, the plant was required to chlorinate this effluent prior to discharge.

2. American Public Works Association, American Society of Civil Engineers, and Water Pollution Control Federation, *Financing and Charges for Wastewater Systems,* Joint Committee Report, 1973.

3. In practice, sampling in Atlanta may occur much less frequently.

3 Sewer Charges and Industry

Faced with a sewer charge, a firm can select one or more of the following responses: pay the fee; pretreat wastes prior to discharge into a sewer leading to a publicly owned treatment plant; improve housekeeping; change process technology; change products manufactured; reduce output, possibly to zero, and go out of business; treat their own wastes prior to discharge into the environment.

This study analyzed the responses at 101 plants in five cities and nine industrial sectors. The first section of this chapter describes the factors that were thought to be critical in determining the response and the procedures for selecting the plants. This is essentially background information, intended to show the potential biases and limitations of the study before the results are described.

The second section describes the actual responses to the charges and documents the major results of the study. That section is organized around a series of fourteen conclusions arising from the work, which show the similarities and differences of responses to wastewater charges in the sample.

Survey Design

The survey design efforts were divided into two tasks: selecting plants to be studied in each of the five cities, and developing a list of information to be obtained during each of the case studies.

Plants were to be selected from a limited number of industrial categories, in order to obtain a sufficient number of plants in individual categories to facilitate generalizations about plant responses within and across categories. All selected plants were in the group currently being charged according to the strength of their wastewater, which included only a small portion of all plants in each of the five cities. The industries selected represented those most frequently using municipal treatment systems and those with significant opportunities for reducing their wasteloads. The categories were meat packing, canning, dairies, other food processing, chemicals, industrial laundries, paper and boxboard manufacturing, metal finishing, and other metal fabrication.

During the initial visits to the five cities, preliminary information was collected on individual plants in these industrial categories. The selection of plants within each category considered the size of the wasteload and charge payments as well as opportunities for reductions. The basic criterion was whether something interesting could be learned about response to charges through the conduct of each case study. For example, if the same product was produced in two cities, plants were selected in both, to provide for the possibility of determining if the responses were affected by difference in the charge programs. The anecdotal information collected from the charge-program staff in the five cities was used extensively in the selection process.

Table 3–1 provides information on the number of plants visited in each industrial category and in each city.

Information was collected on plant economics, production processes, and responses to the charges. The data were collected from plant managers, plant engineers, and financial officers. The categories of information required and the rationale for collecting the data are as follows.

Plant Economics

1. There might be economies of scale in certain of the possible responses to charges, such as pretreatment. Further, reduction in output is itself a possible response.

2. Thriving and dying firms may react differently to a charge. The possibility of exit and failure in response to charges should be higher for marginal operations.

3. A firm's willingness to retire its existing capital stock to install new equipment that is more efficient in resource uses (and generates less waste per unit output) will depend on the age of the existing capital stock.

4. Firms might be reluctant to invest in capital-intensive options for waste reduction (such as pretreatment) if debt financing is particularly expensive for them or if there is limited capital because of other decisions outside the control of the plant manager.

5. The nature of ownership—proprietorship, partnership, closely held corporation, public ownership, wholly owned subsidiary, or branch plant—may affect profit motives, desired public image, and other incentives, which will, in turn, affect responses.

Plant Production

1. Charts were prepared at each plant to give a better understanding of the processes and material flows and to illustrate differences between the plant and others in the same industry category.

Table 3-1
Number of Plants Interviewed by City and Type of Plant

Type of Plant	*Atlanta*	*Chicago*	*Dallas*	*Salem*	*South San Francisco*	*Total*
Canneries	0	1	1	9	0	11
Chemicals	4	9	4	0	2	19
Dairies	2	3	4	0	10	
Other Foods	4	6	5	0	2	17
Commercial laundries	3	0	3	0	1	7
Metal finishing	1	6	2	0	0	9
Meat packing	2	8	4	1	2	17
Other metal	2	0	1	0	0	3
Paper	1	2	4	0	1	8
Total	19	35	28	10	9	101

2. Product types with brand variations were listed. This information helped to identify the industrial affiliation of each plant for analyzing waste characteristics; it also helped to indicate the ease with which plants could shift from one product to another in response to a charge.

3. The composition of inputs also affects the resulting waste stream, therefore, information was collected on the categories, prices, and quantities of inputs. These data were collected for the present and for the precharge period where possible.

4. Time variations in production, including both seasonal and short-term variations (for instance, cleaning on second shift), might alter residuals discharges. Production characteristics affect the set of options plants have in responding to charges. Therefore, data were collected on these characteristics.

Responses to Charges

This set of questions probed the decision-making process that forms the basis of the responses at the individual plants. The responses might include one or more of the following actions:

1. Do nothing (pay charges).
2. Attempt to alter charge levies (request resampling, argue water retention in product, and the like).

3. Commission an engineering study.
4. Change in-plant housekeeping.
5. Make in-plant modifications in production equipment.
6. Pretreat.
7. Change product specifications.
8. Change inputs.
9. Treat and discharge with NPDES permit.
10. Close down.
11. Attempt to modify system (through applications for variances, litigation, nonpayment, and the like).

For each of the responses selected by the plant being studied, cost information was collected on the amount of capital investment, the method of financing the capital investment, annual operation and maintenance (O&M) costs of pretreatment or self-treatment, costs of process or product change alternative methods, and monitoring costs incurred by the plant.

Managers were asked their reasons for rejecting the options for response that were not adopted. Questions were also asked concerning how responses would change if charges doubled in five years in real terms or if they remained constant over five years.

Managers' preferences were probed concerning the use of charge systems rather than standards for achieving environmental goals.

In multi-plant firms, it was asked whether decisions were made at the plant level or at national headquarters.

The effect of other environmental regulations (for instance air and solid waste) on responses to charges was also studied, as was the level of effort devoted to selecting the response to charges. The payback period used in selecting response to charges was also considered.

Survey Results: Industrial Responses to the Charges

This section describes the responses made to the charge-systems at the 101 case-study sites. It is organized around the conclusions reached by the research staff at the end of analysis. Presented is the staff's understanding of how the charges were viewed at the plants and what was done to reduce them. The technique included preparing interview guides, conducting the interviews, and preparing case-study reports from the completed interview guides. These were circulated for review within the project staff, and copies were sent to the people interviewed for their comments and corrections. After the reviews, the case studies were discussed among the staff members responsible for field work. The responses were then categorized, leading to the following conclusions:

Charges are noticed at the plants, and there is a response.

Some responses are administrative only, with no changes in wastewater flows and strength.

Other responses attempt to reduce waste loads and are similar to those made in response to prices of other factors.

There are significant differences in responses in the five cities.

To some degree, plants in the same industry respond similarly.

Responses were small; in only a few cases were the responses large; prices would have to be much higher to achieve large (80 percent) reductions in total effluent volume and strength.

Nonresponse and fraud appear to be infrequent.

There were cases of innovation in the responses.

Conservation measures in wastewater, water, and energy seem generally to reinforce one another, and regulations on discharge components, such as toxics, generally also reduce discharges of BOD and TSS.

Regulations on air pollution, solid waste disposal, pure foods and drugs, and safety seem to lead to larger amounts of discharge and higher charges than would otherwise be paid.

There were significant differences between the responses made by local firms and by national firms producing similar products.

There was limited evidence that the monitoring data from the charges could be used to improve plant operation; however, this potential was not realized, in part because of the low frequency (and sometimes limited quality) of the samples

Plant managers view the charges favorably, as part of the overall cost of production.

Each of these conclusions will be discussed separately, and evidence relating to it will be presented. Before describing the conclusions, however, it is important to summarize the context for them.

The Context for Response

People familiar with the writings of business correspondents, trade associations, or others concerned with the economic health of the country understand that the managers of manufacturing plants and other industrial facilities are subject to numerous regulations that constrain their behavior. The purpose of this study was to investigate plants' responses to charges on wastewater, which generally represent only a small portion of production costs. At the same time, though, many of the plants were required to control discharges of prohibited wastes, to limit their emissions of air pollution, or to control their solid waste disposal.

In the two western cities, plants were forced to reduce their water consumption significantly because of the drought of 1976–1977. In every city, energy prices were described as an extremely important issue in plant operations. Regulations on food production, such as FDA or USDA sanitation requirements, affected water use; so did employee-safety regulations implemented by OSHA. Some changes in water use occurred because of shifts in prices of other factors, and capital availability or technical knowledge affected the responses of some firms.

One of the most significant problems in the study was identifying the effects of the charges themselves within this multitude of other factors. In some cases, savings in the wastewater charges arose as a result of operational changes in response to other factors, and the reverse also occurred. In essence, we have listed changes as "responses to the charges" when there was evidence that there was an intent to reduce charges through the change, albeit sometimes along with other goals. In some cases, these responses may not have been effective—that is, they may not have led to any actual charge reduction, whatever the intent. In other cases, they were clearly effective, as shown by the monitoring records. Because of the limits of these records, however, our major concern has been with intent, not effectiveness. This is the basis for the discussions and tables presented here.

The Charges Are Noticed at the Plants and There Is a Response. A surprising number of plants reported that there was some response to the wastewater strength and flow charges. The charge bills generally represented only a small portion of production costs. In Dallas, for example, 43 percent of those paying charges paid less than $3,000 per year. Only in a few cases did the charge amount to more than a few percent of total operating cost. It is a major finding, then, that these modest costs are of concern to plant managers and that they spend time considering methods for reducing them. There were very few plants in which some sort of response was not at least considered. Some changes were actually implemented at most sites.

These changes are described in the following sections in some detail. Table 3-2 presents the raw data; it lists the case-study plants in each city and industry group and the major responses identified in the case studies by category.

Table 3–2
Case-Study Summaries

				Waste-Reducing Response			*Alter Production*							
Case Study	*City*	*Industry*	*Administrative Response*	*Decrease Water Use*	*Improve House-keeping*	*Install Treatment*	*Equipment*	*Inputs*	*Processes*	*Product*	*Analyzed Problem*	*Innovative Solution*	*Uses Self-Monitoring Data*	*Conserves Other Inputs*
1	CHI	OF				R	R				X			WIE
2	ATL	D	N		RP		SRP							WIE
3	CHI	D			S	R	SRP				X			I
4	ATL	MF	NQ			R		R			X			IE
5	DALL	MF		X	S	SR			S			S		WE
6	DALL	CA	MCNQ	X	S	S								
7	ATL	IL	CN	X			S	S	S					WE
8	CHI	CH	M	X	P	R	SR	SR			X		X	WI
9	CHI	MP		X	RS	SR								WIE
10	ATL	MF	N			R			SR					
11	DALL	D	N		PS		PR							WIE
12	DALL	MP	N											
13	DALL	P		X		PS		S			X			WIE
14	DALL	D	C	X	PS				SP		X			WE
15	DALL	OF	C											
16	DALL	MF	C	X	P	R								WI
17	DALL	D	CN	X	PS	R			S				X	
18	DALL	CH	MN			R					X			WI
19	DALL	CH	MQ			R					X	R		
20	DALL	OF	C	X	PS		SP		SP					WIE

Table 3-2 continued

				Waste-Reducing Response										
							Alter Production							
Case Study	City	Industry	Administrative Response	Decrease Water Use	Improve Housekeeping	Install Treatment	Equipment	Inputs	Processes	Product	Analyzed Problem	Innovative Solution	Uses Self-Monitoring Data	Conserves Other Inputs
21	DALL	D	C	X	SP						X			WIE
22	DALL	OF	NQ	X	S						X			W
23	DALL	CH	MQ	X	SP	R			SP					
24	DALL	OF	N	X		SPR		R				S		WIE
25	CHI	CH		X	PR	R					X	S	X	I
26	CHI	OF	A	X	S						X	X	WI	
27	CHI	D		X	SR		SR				X			WI
28	SO	OF	N	X	S	SR			S		X			W
29	SO	OF												WE
30	CHI	OF			PS						X		X	
31	DALL	IL	C			S								
32	DALL	P		X		S		S			X			WI
33	DALL	CH	NQ		PS						X			WI
34	DALL	IL	N			SP						S		WE
35	DALL	IL	N		S	S	S	S			X	S		WE
36	DALL	OF	CN			SR					X	S		WIE
37	DALL	P	MC			S					X			
38	DALL	MP				SR					X			I
39	DALL	OF	MNQ	X	PS	PS					X	SP		WIE
40	CHI	MP		X	PSR	PSR			SP			S		WI
41	CHI	MP			SPR	R					X	R		WI

42	ATL	IL	N	X			S	S	S		X			WIE
43	CHI	MP			PR	R								I
44	ATL	OF	MN		SP	S	SP							IE
45	CHI	OF		X	SPR	R			SR		X	S		IE
46	ATL	IL	N	X				S						WE
47	DALL	MF	NQ		S	R			SR				X	WI
48	CHI	OF			SPR	R			SR					WE
49	CHI	D			PR				R		X			WIE
50	CHI	D	Q		R							R		WI
51	CHI	CH	Q		PS						X			WIE
52	CHI	CH		X	S				S		X			WIE
53	CHI	MF		X							X			W
54	CHI	MF		X	R	R								WIE
55	ATL	MF	MQ	X	R	SRP					X			WI
56	ATL	OF	N		S	S	SP		S					WI
57	CHI	CH	Q		S	R	SP						X	WIE
58	CHI	CH	Q	X	S	SR						S	X	WI
59	ATL	D	MCNQ	X	PS						X		X	WE
60	CHI	MP				SR					X			
61	CHI	MF	Q	X		R				SR		R	X	WE
62	SF	IL	CQ	X				S			X		X	WE
63	SF	OF	CQ	X	SD		S				X			W
64	SO	OF	M	X		SR	S				X			WIE
65	SO	CA	M	X	S	SR			S		X			WIE
66	SO	OF	M	X		SR								WIE
67	CHI	MF		X		R			R					WIE
68	SO	CA	M	X	S	SR			S					WIE

Table 3–2 continued

				Waste-Reducing Response										
							Alter Production							
Case Study	City	Industry	Administrative Response	Decrease Water Use	Improve House-keeping	Install Treatment	Equipment	Inputs	Processes	Product	Analyzed Problem	Innovative Solution	Uses Self-Monitoring Data	Conserves Other Inputs
69	SO	OF	M	X	S	SR		S			X	S		WIE
70	SO	CA	M	X	S	SR			S			S		WIE
71	SF	OF	CN	X			DS							WIE
72	SF	OF	C	X	S	SR								WIE
73	SF	OF	C	X	S	SR	S	S	S					WIE
74	CHI	MP			SP	R	S							WIE
75	CHI	P		X	S	S					X	S	X	WIE
76	CHI	P				SP								WIE
77	DALL	MP			S	S					X			WIE
78	CHI	MF				R								I
79	CHI	MF				R						R		I
80	ATL	CH	MCQ			S					X			WI
81	ATL	P	MNQ	X		SP	S	S	S				X	WI
82	ATL	CH	MNQ	X	S	R	S		S		X			WI
83	CHI	CH	M		RS	R								W
84	CHI	OF			S									
85	DALL	P	M	X		SP	SP				X			WI
86	ATL	CH	N			S		R			X		X	W
87	SF	P	MC	X	S	S	R	R	R		X	R	X	WI
88	SO	CA	M	X	S	SR					X		X	WI

89	SO	CA	M	X	S	R					X			WI
90	ATL	MP	NQ		S	S	S	S				S		WI
91	SF	MP	C	X		S	R		R				X	W
92	SF	CH	MN	X	RDS	S		R				S	X	W
93	CHI		C		R	R						R		W
94	CHI			X	S							R		W
95	CHI		C	X	R	R								WI
96	ATL		C	X	SP									WI
97	ATL		CN	X	SP									WI
98	ATL	OF	CNQ			S								
99	ATL	CA	CNQ		S						X			
100	CHI	CA	C		S	RS	S			S	X			WIE
101	SF	CH	MCNQ	X	SR	RS					X	R	X	WIE

KEY:

X = Yes
Blank = No or No Data

Cities:
CHI = Chicago
ATL = Atlanta
DALL = Dallás
SF = South San Francisco
SO = Salem, Oregon

Industries:
MF = metal finishing
OF = other foods
D = dairy
CA = canneries
IL = industrial laundry
MP = meat processing
CH = chemicals
P = paper

Administrative Responses:
M = metered effluent flow
C = argued for in-plant water consumption
N = requested new samples to reduce charges
Q = split samples w/district for analysis

Causes of Response:
D = because of drought
S = because of surcharge
R = because of regulation
P = because of product

Other Conservation Actions:
W = water
I = inputs and products
E = energy

As can be seen from the summaries at the bottom of the table, the types of response varied considerably, with many of the changes being made jointly in response to the charges and other factors. However, of the 101 plants visited, only 7 made no response at all, and only 20 did not make a response that was expected to actually reduce their wastewater flow or strength. Given the low level of the charges, this is a surprising level of concern. Some of the reasons for it will be assessed here.

Some Responses Are Administrative Only, with No Changes in Wastewater Flows or Strengths. Administrative responses are categorized as those that resulted in no actual changes in wastewater volume or composition but changed the charges paid on that wastewater. There were several ways in which this could happen; four were predominant.

The first was the ability to request a new sample in some of the cities. If a particular measurement of the wastewater turned out higher than expected—either because of in-plant problems or because of possible measurement error—the plant manager could request that a second sample be taken. The firm would pay the costs of sampling and analysis, but the expected lower reading could save charge payments, resulting in a net gain for the firm. Table 3–3 shows the results achieved from such requested samplings in Dallas. It is obvious that samples are only requested when the expected savings are very large.

A second and related response was to split the sample with the district in order to provide independent checks on the laboratory analysis. Here, the goal was more to make sure that the results were reasonable than to reduce charges directly, although significant differences could result in further action by the firm.

Third, charges were assessed for many plants based on the metered water flowing *into* the plant, either because there were no existing sewer meters or because of perceived problems in them. Since the charges were generally computed by multiplying a rate times the volume of flow, any reduction in the estimated flow would reduce charges. Therefore, many plant managers made the case for water lost in boilers, drying operations, and cooling towers, or for water retained in such products as industrial cleaners mixed up from powders. In some cases these estimates were backed up with data on plant operations; in other cases, rough estimates were used. The level of analytic sophistication varied among the cities and was affected by the level of reduction in charges being sought. It was also possible in some cases to voluntarily monitor effluent flow as an alternative to assessing in-plant water consumption, which is a fourth type of response. Table 3–4 shows the number of plants making each of these characteristic responses within the sample.

Some of the specific cases are interesting. In one, a 1,000-gallon drum

Table 3-3
Results of Extra Samples Requested by Industry in Dallas

	Number	*Percentage*
Net Loss to Plant:		
Higher loadings and higher charges ($1,000–$5,000)	2	18
Savings[a] less than cost to the plant of new sample ($100)	0	0
Net Gain to Plant:		
Savings of $100 to $5,000	3	27
Savings of $5,000 to $10,000	2	18
Savings of $10,000 to $20,000	0	0
Savings of $20,000 to $50,000	3	27
Savings over $50,000	1	9

[a]Projected over six-month period, typical between samples.

Table 3-4
Frequency of Various Types of Administrative Responses

Administrative Response[a]	*Number of Plants*	*Comments*
Metered effluent flow directly	26	Required in ten Salem plants
Negotiated revised estimates of in-plant water consumption	28	This estimate is probably low
Cheated	3	These three sites admitted modifying production during sampling, or other similar actions; others may have done so but did not state this in the interviews
Requested new samples	33	Only possible in South San Francisco, Dallas, and Atlanta (58 plants)
Split samples	25	Most common in Atlanta (9 of 19 plants)

[a]Multiple responses possible.

of corn syrup was spilled while the district was monitoring. Since this was not considered to be usual practice, a new sample was requested (at company expense), and the charges were lowered. In another instance, a laundry was installing a new heat exchanger when the initial sampling took place, and the pretreatment system was disassembled. A new sample was taken after the system was back in operation, and charges were lowered for the remaining months of the sampling period. The manager at one chemical company felt that minor spills and the like were fate and that the sampling results would average out over time to make an accurate record. This shows unusual confidence in the monitoring methods, especially considering the infrequent sampling carried out in most cities.

At several sites where the district is responsible for monitoring, plant managers would request a portion of the sample for separate analysis. In one case, where the strength-related charge was extremely high ($250,000 per year for a plant employing thirty-five people), this was routine practice, although little was done with the results. Managers at another large plant, however, were sufficiently bothered by the results to have an outside assessment of the district's laboratory performed, and they continue to be concerned with quality control in the measurements.

In some cases, managers were willing to invest relatively large sums to measure water retention inside the plant or to monitor wastewater flows. One small chemical mixer installed submeters at a number of locations so that the water entering the product could be accurately measured and subtracted from the bill. At other plants, engineering studies were performed to calculate estimated boiler losses and to assess the amount of water evaporated from clothes dryers. True water conservation was also an important response, since it generally affected the volume-related charges, at least, and these were often large.

Other Responses Attempt to Reduce Waste Loads. There were few differences in the ways plant managers treated wastewater charges in comparison to changes in the costs of other factors of production. There were a few attempts at political or judicial action to reduce the bills, but these were the exception. Most actions sought to actually reduce the charges by reducing waste load, or to get better measurement (at least, more favorable measurement), as described earlier. Table 3-5 classifies these responses into a number of types, as described here.

Table 3-5
Frequency of Various Types of Physical Responses

Physical Responses[a]	*Number of Plants*	*Comments*
Water conservation	38	Strong drought response in Salem and South San Francisco
Improved housekeeping	53	Most common response to charges
Installed or augmented treatment	44	In many of these cases, regulations were the main reason for installing pretreatment
Altered production	41	Includes changing products, inputs, process flow, or equipment

[a] Multiple responses possible.

Installed or Augmented Pretreatment. Plants in this category installed some form of pretreatment, ranging from settling basins or screens to vacuum filters or biological-treatment facilities. Relatively few of these installations were responses to the charges alone, but charges were considered in designing the pretreatment systems at a number of sites. The pretreatment systems required by regulations were also improved to reduce charges, through technology changes such as the addition of finer screens or through increased attention to operation. However, most of the more effective and more expensive pretreatment systems were introduced in response to regulations controlling the discharge of particular substances. A few, though, were charge-related, for example:

> An industrial laundry, which built a settling tank and oil skimmer to partially clean its wastewater before discharge.
>
> A boxboard producer, which installed a holding tank downstream from its corrugator to capture some of the solids.
>
> A fats and oils producer, which installed a dissolved air flotation system to recover product and reduce wastewater charges, with both savings required to make the system a good investment.

A number of examples of significant pretreatment in response to charges occurred in the paper industry, which will be discussed later.

Process Improvements and Product Shifts. One of the most significant types of changes made by a number of plants was the modification of production processes, changing inputs or outputs to reduce charges.

> A large bakery sells its waste products rather than sending it to the sewer.
>
> A number of canneries now landfill their peelings rather than flushing them.
>
> An engineer/builder has begun to reclaim its cleaning solutions in response to regulations as well as surcharges.
>
> Several paint and chemical companies invest considerable resources in scheduling to minimize cleanup and tank washings.
>
> Dairies frequently run products with increasing amounts of fats as the day wears on to make use of fat buildup.

Other common responses were to change equipment to reduce product losses and to lower wastewater loadings.

> A margarine and oils producer, a chicken processor, a meat packer, and two dairies all installed better processing equipment to reduce spills.
>
> Three paper companies improved their internal screening system to enable them to reuse more water and pulp.
>
> Some canneries installed dry conveyance for their products to reduce water use, wastewater flow, and waste loads.

Many of the canneries, as well as some other companies, mentioned that, although wastewater production was not a major factor in the selection of new processing equipment at current rates, it was considered. At one catalyst manufacturer, for example, it was claimed that wastewater control was used as a partial justification for the replacement of older, less-efficient equipment.

Sometimes, however, wastewater improvements are a side benefit. One chemical plant has made a steady effort to improve its production efficiency by rearranging production equipment and adding plumbing to replace manual chemical movement. These resulted in lower solids loadings and less wastewater use.

The responses of laundries varied.

> A number of laundries installed dry-cleaning equipment for products such as oil mops in order to lower TSS and other waste components.
>
> Other laundries eliminated certain processes or products.
>
> One large laundry cut out its commercial linen service because it raised the solids concentration excessively.
>
> Another dropped its "dirtier" customers because of discharge limits on lead and other constituents in response to regulations.

In a similar vein, two paper manufacturers eliminated the use of lead inks because of effluent limits, and a cannery explained to its growers the importance of minimizing the amount of dirt included with the incoming vegetables. Still others have altered their raw materials and their production processes.

> A meat company decreased the amount of frozen products used in order to reduce the amount of defrosting needed.
>
> A chicken processor stopped slaughtering the birds at the plant.
>
> A paint company discontinued a few products entirely, but this response was relatively infrequent.

However, other-factor prices have adversely affected wastewater quality through a similar mechanism. For example, one large meat-processing firm is about to start buying frozen hams, which will greatly increase fats, oils, grease, and BOD loadings but is expected to be more profitable overall.

The production engineer of a detergent manufacturer explained that waste generation could be cut by 10 percent if the plant did not produce so many different products. The different soaps and detergents produced have similar chemical make-ups, differing primarily in scent and color but requiring thorough equipment washes between batches.

Housekeeping and Operations. For much of the food industry, for the dairies, and for some chemical plants, improving housekeeping represents the most frequent response to the charges. Improvements were achieved chiefly through employee training: explaining the charges and the methods for minimizing them, careful recovery of spilled imperfect product, and dry cleaning prior to washdown of equipment and floor. One manager of a a dairy stated that he believed in a dry-floor operation and would impress this on the staff, to the extent of discharging employees who showed a lack of concern. Product savings often resulted in addition to the surcharge savings. One large meat processor supplied its employees with proper equipment and provided annual training sessions to maintain high standards of housekeeping.

One large chemical company began a concentrated campaign to reduce its solid loadings through improved product recovery. Catalysts are produced using a slurry suspension, which is subsequently filtered and then dried into pellets. The filters can rip, allowing the chemicals to escape. Furthermore, there are always some chemicals lost in the drying process. By employing settling tanks and emptying them conscientiously, and by keeping careful watch over the filters, the plant is able both to save product and to lower its surcharge. Prior to the recent increases in charge levels, the company was much less careful. Now they use daily solids-concentrations data to spot problems in their production.

Another chemical company is in the business of reclaiming valuable chemicals from the wastes of other industrial concerns. The process is primarily fractional distillation, with some liquid/liquid extraction. The company is careful about minimizing product loss and thus about not letting any material into the sewers. But they also found that the charges have made them still more concerned; for example, in response to the charges they have raised all storage tanks off the floor so that catch pans can be placed underneath each container. More time is now spent seeing that hoses are properly connected and that valves and gaskets do not leak.

A margarine producer responded to product loss and the charges. Among other things, all floor drains were closed to prevent major or minor spills of oil or product from passing into the sewers. The first rinse from the equipment-cleaning operations is stored in catch basins to allow the oils to rise to the surface overnight. Margarine is recovered from the improperly packaged individual containers of all sizes, using a hot-water bath and settling tank. Further, they have installed drip pans under the processing equipment to recover wastes, and they pump their grease trap weekly.

It was also found that surcharges provided an incentive for plant managers to run their treatment equipment well, particularly those who had originally installed systems because of regulations.

> An aluminum anodizer claimed that, because of the charge, he was very careful about the proper operation of his system. The company was reportedly careful to replace clogged or otherwise disfunctional pumps. City data showed the solids loadings per unit of production had dropped continually.
>
> A meat-packing company had a large Imhoff tank and skimming system, inherited from the days when it was a rendering operation. It has subsequently stopped rendering but maintains the equipment to reduce charges.
>
> A small paint company pumps its settling tank more frequently than it did before increases in the charge levels.
>
> Since a recent increase, an engine-rebuilding plant is reportedly more careful about pumping one of its settling tanks after cleaning many engine parts.

At numerous sites, plants had contracted with disposal services to pump their tanks routinely, which avoided overflows and reduced total charges (including wastewater and solid waste).

Some plants changed their housekeeping and production procedures because of the regulations. In Chicago, several dairies instituted rigorous housekeeping procedures to control fats, oils, and greases, and the surcharge savings were purely incidental. As one dairy manager said, "I don't care what I pay in surcharges as long as I do not violate the fats, oils, and grease standard." Similarly, meat packagers and processors, as well as food processors, frequently were careful about housekeeping because of the standard.

Decreased Water Use. Although the primary focus of this study was on the reduction of wastes entering the sewers, reductions in water use also affected total wastewater charges and were extremely common. They ranged from internal recirculation of cooling water to simple installation of automatic shutoffs on hoses. Frequently, these measures also reduced strength-related wastewater charges by generating solid waste instead of liquid. For example, a number of canneries replaced their wet conveyance system with conveyor belts, both decreasing water use and preventing waste fruit from entering the sewers.

Water conservation was most important in Salem and South San Francisco, in response to the recent drought. In some of the other cities, however, increasing wastewater charges often provided an impetus for water conservation. This cross-elasticity might be an important result.

Summary. Many of the individual responses described in this section represent examples of the range of response found in the case studies. There is a great deal of variation in the type, magnitude, and effectiveness of the responses. This diversity is reasonable, given the varying regulatory context, charge levels, and types of plants in question.

There Are Significant Differences in Responses in the Five Cities. Table 3-6 shows the pattern of waste-reducing responses in each city. As previously mentioned, Salem and South San Francisco were affected by the 1976–1977 drought and predated conservation efforts: water conservation was much more prevalent in these cities than elsewhere. Chicago and South San Francisco had strong prohibited-waste enforcement programs as compared to the other cities—particularly Chicago's regulation for fats, oils, and grease. One result was more extensive pretreatment than in the less-controlled areas, such as Dallas and Atlanta. Similarly, Salem set minimum pretreatment standards for canneries; the standards were implemented.

Table 3-6
Waste-Reducing Responses by City
(*percentages*)

City	*Decreased Water Use*	*Improved Housekeeping*		*Pretreatment*		*Production Changes*	
		Total	*Charge-Related*[a]	*Total*	*Charge-Related*[a]	*Total*	*Charge-Related*[a]
Atlanta (n = 19)	47	53	42	58	42	74	53
Chicago (n = 35)	46	77	54	71	23	37	26
Dallas (n = 28)	50	50	46	64	54	43	39
Salem (n = 10)	100[b]	70	70	90	80	60	60
South San Francisco (n = 9)	100[b]	67	67	78	78	67	44

[a]Percentage of total plants in city, not of all those responding; included even if charge was only part of the reason for the response.

[b]Response to drought.

In Dallas and Atlanta, where the charges acted as the incentive for reducing conventional pollutants, the result was less-extensive treatment and more variation in response.

Pretreatment was implemented frequently in response to regulations, with operational changes more likely to be caused by the charges. This picture would be even clearer except for one administrative response to Chicago's regulation on fats, oils, and grease. Dairies in Chicago contested the regulation, arguing successfully that butterfat was beneficial to the treatment works and should be excluded from the test. As a result, almost all dairies were able to get by with extensive housekeeping and did not need to install pretreatment.

On the face of it, this pattern of variable operational responses to charges, and pretreatment responses to regulations, appears to be an interesting result that supports the concept of effluent charges. However, the reason for pretreatment in these cases was not usually that a particular technology was required (for instance, BAT, NSPS) but rather that the effluent standards were hard to meet. These standards were performance based, except for the screens required in Salem, which would also have been efficient responses to the wastewater charge, and were sometimes difficult to achieve, while the charges were generally low, not warranting major changes. All five cities had both standards and charge-systems; however, the relative importance of each varied to some extent. For example, Dallas and Atlanta had the same standard for fats, oils, and grease as Chicago, but enforcement in the two southern cities was much weaker. Thus, there was less overall response from the dairies, meat packers, and oil producers.

To Some Degree, Plants in the Same Industry Respond Similarly. Table 3-7 shows the patterns of response for the "pure" industries in the sample—ignoring the "other" categories and also excluding the canneries, since they were almost all in one city. Discussions of the remaining industries follow.

Paper. Two types of plants were generally included in the survey: producers of paper products from wastepaper and producers of corrugated boxes from purchased paper. All the paper producers were sensitive to the waste charges as a significant cost, and all used high-technology methods to reduce their waste loads.

These generally involved increased screening of effluents, more internal recycling, and some consideration of zero discharge if the charge rates rose significantly. There were attempts to conserve both product and water; however, existing equipment and processing procedures, particularly in areas such as storage capacity, reduced the effectiveness of these efforts in some plants. Charges were much lower at the boxboard plants. Correspond-

Table 3-7
Characteristics of Response by Industries
(percentages)

Industry	Decreased Water Use	Improved Housekeeping		Pretreatment		Altered Production	
		Total	Charge-Related	Total	Charge-Related	Total	Charge-Related
Paper ($n = 8$)	75	25	25[a]	100	100	62	50
Dairy ($n = 11$)	55	100	73	18	9	73	45
Meat processing ($n = 12$)	25	67	50	92	58	25	17
Metal finishing ($n = 12$)	58	42	17	83	17	42	25
Commercial laundry ($n = 7$)	57	14	14	42	42	71	71
Chemicals ($n = 19$)	53	74	58	68	26	42	21

[a] 25 percent here means that two of the eight plants made charge-related housekeeping improvements or that 100 percent of all housekeeping improvements in the paper industry were at least partially charge-related.

ingly, the responses were limited to installation of settling tanks and to operational changes. These plants, though, were subject to lead-effluent limits from their inks, and they found at least two solutions: changing inks and reusing washwater.

Dairies. The major problems in the dairies were disposal of acid whey from cottage cheese, disposal of returned milk, plant washwater, and wastewater from equipment cleaning. Returned milk and other spoilage were uniformly sold or reused in some products. The whey washed from cottage cheese remained a disposal problem at almost all plants, and most whey was dumped into the sewers. A number of plant managers were actively concerned with this issue and attempted to recycle washwater, sell the whey or give it away, or dispose of the waste through scavengers. The costs of recycling, at this point, were higher than the charges incurred by sending the whey to the sewers. Plant washwater was another continuing concern, and plant managers constantly attemped to reduce water use and wastewater flow. These attempts generally involved flow cutoffs or hoses, reduced filling speeds where possible, and some mopping before washdown. Almost universally, the wastewater charges were viewed as an incentive in this area, although water conservation and efficient operations were the real goals of the changes.

A number of approaches were used to minimize cleaning of tanks, at least into the sewers. Several plants had recycling cleaned-in-place (CIP) systems; others air-flushed lines to recover products between batches before cleaning with water; others increased fat content during the day to reduce the need for cleaning build-up. Responses in this area showed diversity.

Meat and Poultry Processing. An often-repeated story in the interviews was that the profit in meat packing comes from the byproducts, not from the meat. Blood, hides, feathers, hoofs, fat, and so forth, are generally recovered for pharamaceutical, tanning, rendering, and other uses. Waste product, which generally includes the sludge and scum from settling or screening, is sold for feed or rendering. These byproduct streams are important, although sometimes they are discovered in an attempt to reduce charges for wastewater disposal. Screening and settling of wastewater are common. However, at least one Chicago plant used chemical floccculants, which increased removal but made byproduct sale or reuse impossible.

Metal Finishing. Control of pH levels was almost universal among the electroplating plants. In addition, there were common methods for reducing discharges, such as the ion exchange of solutions. In general, though, relatively little was done in response to the charges because they were insignificant in comparison to the regulations at all sites. The only exceptions were a few anodizers with high TSS but relatively nonhazardous discharges. At one plant, for example, aluminum sulfate was recovered in large quantities and used as a building material. Before installation and operation of pretreatment, it led to both high charges and blocked sewers.

Laundries. The industrial laundries showed three typical responses, which were usually cumulative. First, screens or settling tanks were installed to capture fibers in the wastewater. Second, some of the dirtier objects, such as oil mops, were dry cleaned and generally recycled. Finally, there were some active attempts at controlling inputs, either by limiting the types of wash accepted or by talking the problem out. At one site, for example, a plant manager apparently convinced a customer not to dirty rags and mops as extensively. The last two responses were usually initiated by existing or potential regulations on substances and not by charges.

Paint Manufacturers and Chemical Producers. This industry showed extremely varied responses, with few obvious patterns. They were generally concerned with cleanup operations, where most of their discharges occurred. A number of them produced thousands of different products and package sizes, resulting in a concern for scheduling. Some of the discharges included prohibited substances, so a number of plants had to install wastewater-flow controls, buffering operations, and the like.

In Only a Few Cases Were the Responses Extremely Large; Prices Would Have to Be Much Higher to Achieve Large Reductions in Total Wastewater Strength. The levels of treatment for industrial pollutants contained in the best-practical-technology (BPT) standards are almost uniformly greater than 50 percent removal; best available technology (BAT) levels are much higher, and the within-system effluent limitations imposed by the cities ranged up to better than 90 percent removal for many pollutants. We believe that the reductions shown in the removal of BOD, COD, and TSS in response to the charges were generally 0 to 40 percent, with only a few fortuitous reductions of greater magnitude. It is clear that much larger charges would have to be imposed to achieve reductions at the BAT level, although there is little evidence on what the charge levels would have to be.

On the other hand, both plant managers and officials at the districts believed that the charges covered the cost of treatment, and they saw little reason to increase the charge levels unless treatment costs rose. It appears that there are economies of scale in plant operations. Issues relating to the incentive properties of the charges for reducing the need for plant expansions have not arisen in these cities.

Nonresponse and Fraud Appear to Be Infrequent. As described in a previous section, only a few plants implemented no physical or administrative responses. Even among these plants, several studied the charges and concluded that "no response" represented the least-cost alternative. Seven plants completely ignored the charges. Four of these were platers in Chicago who had significant pretreatment to meet standards.

The plants generally appeared to work within the system. We have one reported case of a bribe—which was eliminated with new plant management. In a number of cases, plants "cleaned up" when the sampling crew was on-site, but these plants appeared to be fairly well known to district personnel and were controlled by spot checks or longer sampling periods. For example, in one case where the district believed the plant was not discharging, samples were kept for several days and then public health inspectors were called in. They found unsanitary conditions for a food-processing facility and shut the plant down, requiring immediate cleanup—with the sampling crew still monitoring the wastewater.

At another plant, an automatic sampler was left on-site for over a month, taking daily composites. The Thanksgiving Day sample showed two to three times the concentrations of other samples, and entered into the rate concentration. Spot checks are also used, apparently with some effectiveness, on the most significant violators. Many managers claimed that they did not know when sampling occurred and that they did not try to make special efforts to reduce discharges during sampling. Both of these claims were disputed by the districts. In any case, it appeared that there were few significant attempts to violate the spirit of the charges, and most were unsuccessful according to both sides.

There Were Cases of Innovation in the Responses. One of the most striking results of the case studies was the level of tinkering with the equipment—attempts to reduce charges or meet effluent standards through variations in approaches, operating methods, scheduling, chemicals, and so on. The plants studied seemed to have adequate levels of operator training and generally high interest in making the systems work. We defined *innovation* as going one step further—actively attempting to find new ways for reducing wastes. Given that definition, several cases qualified.

Perhaps the most clever innovation was developed at a corrugated box production facility. The plant engineers experimented with using the flexographic press washdown water, high in heavy metals, to make up the starch emulsion used to corrugate the cardboard. The process was successful and the plant no longer has an effluent problem.

A large private dairy researched the possibility of building a commonly owned whey-recycling plant, but the project was dropped because of cost considerations and fear of antitrust action. Similarly, a group of canneries investigated developing common spray irrigation systems to lower their waste loads. This was dropped because of its expense.

A shrimp-processing plant was able to hire an engineer who, a number of years previously, had developed a screening and recirculation system for its wastewater. At the time, there was no formula solution to their problem. A small poultry processing plant used an old feather separator to lower solid concentration in its effluent. A frozen-vegetable plant used its bean washwater and its cooling water for spray irrigation. A large national canner installed a larger conveyor belt to minimize the amount of wastes falling to the floor.

One food-oil producer designed a dissolved air flotation system, which will pay for itself by lowering the company's surcharge payments by 40 percent, improving product recovery of waste oil, and lowering water use. This system represents a variation or refinement of a system proposed by a consultant and field-tested at the plant. The new design is expected to remove 75 percent of the waste oils at 10 percent of the cost.

Conservation Measures in Wastewater, Water, and Energy Seem Generally to Reinforce One Another, and Regulations on Discharge Components, Such as Toxics, Generally also Reduce Discharges of BOD, and TSS. In this and the following two sections, we are concerned with how the context of these plants complements or adversely affects the responses to wastewater charges. Our concern here is with conservation and effluent-control measures; the following sections deal with other regulations and consider organizational structure.

The first conclusion is that most plant managers appear to be active cost minimizers. Most plant managers expressed concern with energy use, product loss, water use, and wastewater strength. When they were actively concerned, they could achieve significant savings: one plant manager doubled productivity and total output in two years and successfully made the case to corporate management that an investment in screening and settling would show high returns. Another reduced water use in an old dairy by 40 percent and improved quality control at the same time. In the West Coast cities, and in a number of other plants elsewhere, it was clear that attention to water conservation also meant reductions in wastewater flows and, at times, strengths. Several such cases have already been discussed, involving recirculation, dry conveyance, and the like. In the early 1970s, a major soap and detergent plant invested in major plumbing improvements, which increased recirculation and rationalized discharges, making significant amounts of product recovery possible. Water and energy were also saved, and operations were improved because of reduced opportunity for error.

Energy conservation has also helped to reduce wastewater discharges in a number of cases. Wastewater heat reclaimers in laundries usually require relatively clean input, implying the need for screens, which in turn reduced wastewater loading. Similarly, recycling of whey washwater in a dairy involved moving chillers much closer to their point of use, at the same time saving energy and water and reducing discharges.

If a product is lost in the wastewater, removing that product generally reduces waste loadings. Waste control at its best results in salable byproducts. There were numerous examples of these. Given the state of the art of waste control, it is also usually true that treatment processes that remove prohibited wastes will also remove compatible wastes, to some degree at least. However, we did have one counter-example, a chemical plant that had a high level of phenol in its discharge. The plant's treatment system consisted of a secondary biological-treatment process, using bacteria to digest the phenol. The result was a nonhazardous sludge, which could be most economically disposed to the sewers.

There are cases in which the interactions between the regulatory and charge systems are not productive, although these are not frequent. Some of the chemical-treatment processes used yielded sludges that could not be used as a byproduct and that required incineration. Most efficiency improvements and discharge regulations complemented charges and helped to reduce discharge volumes and strengths.

Regulations on Air Pollution, Solid-Waste Disposal, Pure Foods and Drugs, and Safety Seem to Lead to Larger Amounts of Discharge and Higher Charges than Would Otherwise Be Paid. The previous section demonstrated that the various efficiency improvements and wastewater controls were usually complementary to charges. This is not the case with other regulations, as shown in a number of examples:

A joint whey-treatment facility proposed in one city was rejected because of antitrust concerns.

A wet scrubber required for air-pollution control resulted in large quantities of waste, prohibited from discharge to the sewers.

Recycling of transport water was impossible in a shrimp plant because of possible contamination (even recycling of just the water used to carry away shells and discards).

Federal public-health standards also required a *minimum* water flow per bird processed in a poultry plant.

Plant officials of a margarine producer were required to wash the floor more frequently because of OSHA concern for slipperiness.

Several plant managers stated that solid-waste-disposal regulations had reduced the frequency with which sludge was pumped, probably leading to increased discharges.

There Were Significant Differences between the Responses Made by Local Firms and by National Firms Producing Similar Products. National firms were characterized by centralized capital budgeting with detailed documentation; a short payback period (one to five years, depending on the firm); rotating plant managers; central engineering staff concerned with wastewater; and possibly some desire to act ahead of regulations and install better treatment facilities. Local firms had less technical experience available; apparently easier capital availability; and less-formal decision-making process.

Although local firms were often supported by trade associations and other industry groups, the national firms shared many advantages:

One firm, in a situation of limited capital availability, was able to set up an equipment trading program among its dairies.

Another firm introduced a treatment technology nationally, in advance of tougher regulations in the district.

Several firms had national engineering departments, which were familiar with wastewater issues at all plants and able to consider such

> options as regional cottage cheese plants to reduce the acid whey-disposal burden on the local dairies.
>
> Shifting plant managers from site to site appeared to result in operational improvements, often copied from previous locations.

In general, the national firms appeared to act more slowly, but with more information.

There were a few cases of generally recalcitrant national firms that attempted to avoid regulations and charges and responded as slowly as possible. However, at least an equal number of local firms showed similar behavior. The national firms, however, appeared to be somewhat more effective in their delays.

In summary, it was clear that the best of the national firms had considerably more information available to them in their decision-making than the local firms, although there was also more competition for investment capital.

There Was Limited Evidence that the Monitoring Data from the Charges Could Be Used to Improve Plant Operation; However, This Potential Was Not Realized, in Part Because of the Low Frequency (and Sometimes Limited Quality) of the Samples. The quality of the monitoring system is fundamental to the effectiveness of the surcharge system. Without good information on wastewater concentrations and flows, billing is inaccurate. If plant managers think that the monitoring data are not representative and that the surcharge bills do not reflect their waste characteristics, they have little reason to respond by lowering their waste loads. If the data vary in an apparently random fashion, they have no basis for their actions. If the plant managers received the results long after the monitoring took place, they might be unable to pinpoint the reasons for a high or low reading.

Monitoring information is potentially valuable to plant managers as well. Monitoring data could be useful in the control of production processes: sampling results can help locate the sources of waste loads, identify product losses, and enable production managers to correct the problem. However, few plants regularly sample their wastes or use the district's samples for control.

In several instances, feedback from surcharge monitoring had an impact on plant operations.

> One chemical company monitors its solids concentration daily. When levels are unusually high, the source of the solids is searched out in order to solve the problem. This reduces product loss and at the same time keeps the surcharge low.

> A powdered soup and soft drink plant, a paint producer, and a corn mill also use monitoring data to guard against unnecessary product loss and to control production quality.
>
> A meat processor has pH monitors in order to run its pretreatment system effectively, and a laundry is considering installing one for process control.

Aside from these plants, however, few do more monitoring than is required by city regulations, and very few use those results for anything.

The question is why, and the answer is fairly simple: the monitoring records are not good enough for process use. They are infrequent (one to four times yearly, except for Salem), and the response time is slow (five-day BOD is tested, except in South San Francisco). Although a few plants can use such data, most cannot. Given the programs analyzed, the potential value of wastewater monitoring to the plants cannot yet be realized.

Plant Managers View the Charges Favorably as Part of the Overall Cost of Production. Voluntary compliance is a necessary ingredient for regulatory systems because no agency has the resources to monitor constantly or to take enforcement actions against large numbers of violators. Voluntary compliance depends to some extent on the perceptions associated with the regulations; rules viewed as fair and equitable are generally well accepted. Sewer charges are perceived as fees for waste-disposal services, which are generally considered a reasonable part of doing business.

Aside from the perceived lack of accountability in the rate-setting process, almost everyone liked the charge systems and preferred the charges to other potential forms of control, such as regulations. The reasons varied, as shown in table 3-8. A few claimed it to be economically efficient for society. More claimed it was efficient for them and wished that the city would run a surcharge system for metals or hazardous wastes (although they had little idea of the cost of centralized removal). One chemical company actually petitioned the city to set up a hazardous-waste surcharge system. Others claimed that the city could hire professionals to run treatment systems, whereas they were only amateurs and, as such, likely to do a poor job. Some claimed that pollution-control regulations were too tied to technology and often too strict. Many just expressed a preference for the system, without any specific reason. Managers of two plants wished that there were pretreatment regulations tailored exactly to each industry but believed this would be impossible to do. Almost every plant viewed the systems as equitable or, at worse, a necessary evil.

It does appear that, in most instances, plant management prefers surcharges to quotas, technology-based standards, or limits on effluent con-

Table 3–8
Preferences for Sewer Surcharge over Standards as a Regulatory Mechanism

Prefers surcharge to standard	49[a]
No reason offered	21
Surcharges allow firm to base waste-management decisions on economic analysis	9
Compliance with standards would be more expensive	3
Surcharges allow firms to have variable discharges	3
Surcharges provide continuous incentive for proper waste-management practices	8
More economical for society	2
Surcharges are equitable	7
Prefers standards	2
Would prefer whichever method costs firm least	3
Indifferent	4
No data	41

[a] Numbers do not add to 49 because some firms gave more than one reason.

centrations. Although most of the reasons cited for the preference for charges were straightforward, a few interviews did mention the incentive properties. However, several opposed general effluent taxes because they did not want to pay for the "right to pollute." They were happy to pay a fair cost of doing business, to cover the district's treatment cost, but were against generalized pricing approaches implemented for their incentive properties alone.

4 Conclusions

At the inception of this study of local wastewater charges, it was clear that there were two major areas for study: how well the charge systems worked and how the industries responded, if at all, to the incentive for waste reduction represented by the charges. Briefly, our conclusions are that the charge systems work well, as viewed by their operators and customers, and that almost all plants showed some response to the incentive properties.

System Operation

Monitoring and billing procedures, rates, rate-setting procedures, related discharge regulations, methods for resolving disputes, and other characteristics all varied from site to site. Administrative cost for strength-related changes depended on many things and ranged from 7 percent to 21 percent of program revenues. Rates were usually set to cover the costs of treating the wastes from the plants, and revenues came close to covering these costs, based on our best estimates.

The major axis for differentiating the operation of the systems was the degree of agency involvement in monitoring. In none of the systems studied did the monitoring approach involve continuous sampling and analysis, although Salem required the continuous drawing of samples. In four of the systems, the local agency took responsibility for drawing and analyzing samples on an intermittent basis—for example, by sampling continuously for a week, twice annually. While there are problems with this approach in terms of estimating accuracy, it is cost effective and generally well accepted in this field. It is subject to some possible bias—particularly when visits are announced in advance or when after-the-fact variances are available; however, significant distortions appear to be rare.

One benefit of accruing to the systems with agency monitoring became apparent—enhanced capability for toxics management. Where monitoring of prohibited discharges was combined with monitoring to set wastewater charges, there was some tendency to go easy on customers with minor regulatory violations. However, financing the combined monitoring program was relatively easy. Further, the capabilities to sample problem sites carefully and to trace plant upset quickly were enhanced. We believe the net effect was beneficial to toxics management.

In the five systems, only a small portion of all industrial plants were monitored for pollutant strength; the decision was usually made on cost-effectiveness grounds, relating expected charge revenues to sampling costs, although without much explicit analysis. We would expect this result to hold true in other areas and for other media.

The analyses of systems are described in more detail in chapter 2 of this book, and agency case studies are in part II. The major conclusion is clear: the charge systems work from the perspective of the agencies involved, the plant managers who pay the charges, and outside evaluators, such as ourselves. From the outside perspective, much could be improved: rates probably do not closely reflect marginal or even average costs; the monitoring programs give only a questionable estimate of actual waste strengths and flows; information often moves slowly between systems and customers; and there are opportunities for cheating and some downward bias in the estimates. However, these are minor in relation to the strengths of the systems and to their accomplishments in both allocation and setting incentives for waste reduction. Relatively simple systems such as these can perform well, without excessive administrative costs.

Incentive Effects

The major conclusions of the case studies of 101 plants in nine industry groups can be broken down into six categories:

1. The basic conclusion is that charges are noticed and that almost all plants respond. Most reduce waste discharges or flow, often significantly (over 10 percent); plant managers actively looked for ways to reduce the bills.
2. The charges for waste disposal and treatment were low, and the responses were correspondingly small; this makes the nearly universal attempts to respond even more surprising.
3. Context was extremely important in determining the actual responses: plant characteristics, regulations, structure and operation of the charges, and management structure of the firm all affected how a plant would respond.
4. Within that context, other conservation efforts generally complemented the waste-reduction effort; regulations on other forms of pollution, product quality, and safety generally interfered.
5. The major lost opportunity was the inability to use the monitoring data in plant operations. This was a result of monitoring frequency and delays in the analysis.
6. Plant managers viewed the charges as giving them added flexibility

compared to discharge limitations, and they generally supported the system.

These conclusions are supported in chapter 3. They show that the programs have the desired incentive effect and are acceptable to the industrial users.

Implications

Based on the findings and conclusions of this study, we are confident that economic incentives work well and can be effectively administered. We believe that similar systems could be developed and implemented for other media, such as air emissions or solid waste. For wastewater, the programs worked well for BOD, COD, and TSS because these substances are accepted indicators of waste strength, there are relatively straightforward procedures for monitoring them, and treatment costs can be attributed to them. Rate setting and incentives would have been much more problematic with phenols or lead, for example. For air, local economic-incentive programs would therefore appear workable for emissions of TSP, reactive hydrocarbons, and other pollutants, with some concentration of sources but no single dominant producer. For solid waste, rate setting by level of implied hazard could work, but illicit dumping presents a difficult disposal problem, despite permit systems. Still, we believe that most of the results of this analysis can be transferred to other cities and other media, and that local economic incentives for pollution control can both be effective in reducing residuals and be easily administered by the designated agency.

Part II: Charge-System Case Studies

This part reviews the industrial sewer charge and surcharge programs administered by five cities: Atlanta, Chicago, Dallas, Salem, and South San Francisco. In four cities, the agency responsible for the collection and treatment activities is governed by city ordinances, while the fifth, Chicago, is a special district. All, however, are self-financing, and the municipal agencies have entered into limited contractual arrangements with neighboring areas for the provision or receipt of waste-treatment services.

The five systems were summarized in part I. This part provides more detail on each of the systems.

5 Atlanta

Agency Overview

In Atlanta, wastewater collection and treatment services are provided by the Department of Environment and Streets, a part of the city government. In addition to providing these services, the department is responsible for the construction and maintenance of highways and streets, traffic engineering, engineering and construction services, sanitary services, and water supply. These responsibilities are divided among the department's six bureaus, as shown in the organization chart in figure 5–1. Wastewater treatment is the responsibility of the Bureau of Pollution Control in the Department of Environment and Streets. Within the Bureau of Pollution Control, the Division of Inspection and Monitoring is responsible for the treatment of industrial wastes and the administration of the surcharge program.

The Department of Environment and Streets was formed in 1974. Prior to that time, a Department of Public Works was responsible for waste treatment, collection, and other related services. The Bureau of Pollution Control was formed in 1957, under the Public Works Department, and transferred intact to the Department of Environment and Streets in 1974.

The Bureau of Pollution Control operates five waste-treatment facilities. These treat liquid wastes from the city of Atlanta and also contract to treat liquid wastes from neighboring DeKalb and Fulton counties. Minor amounts of effluent from the city of Atlanta are also treated in facilities located in Cobb and Fulton counties and operated by these counties. A map of the service area is presented in figure 5–2. Most of the effluents from the city are discharged into the Chattahoochie River, but, until a new project is completed, lesser amounts are also being discharged to Utoy Creek, Intrenchment Creek, South River, and Flint River.

At its formation, the Bureau of Pollution Control stated its goal to be the effective and efficient treatment of wastes discharged by the city and other contributing areas. This long-term goal has not been changed, but a short-term goal of meeting permit requirements under the NPDES program has been added. Bureau representatives feel that the long-term goal has been achieved, but the city has had recurring permit violations because of breakdowns during the period after its Clayton Plant was expanded.

The Bureau of Pollution Control and the Bureau of Water Supply have combined finances. Billing procedures are the main reason for this:

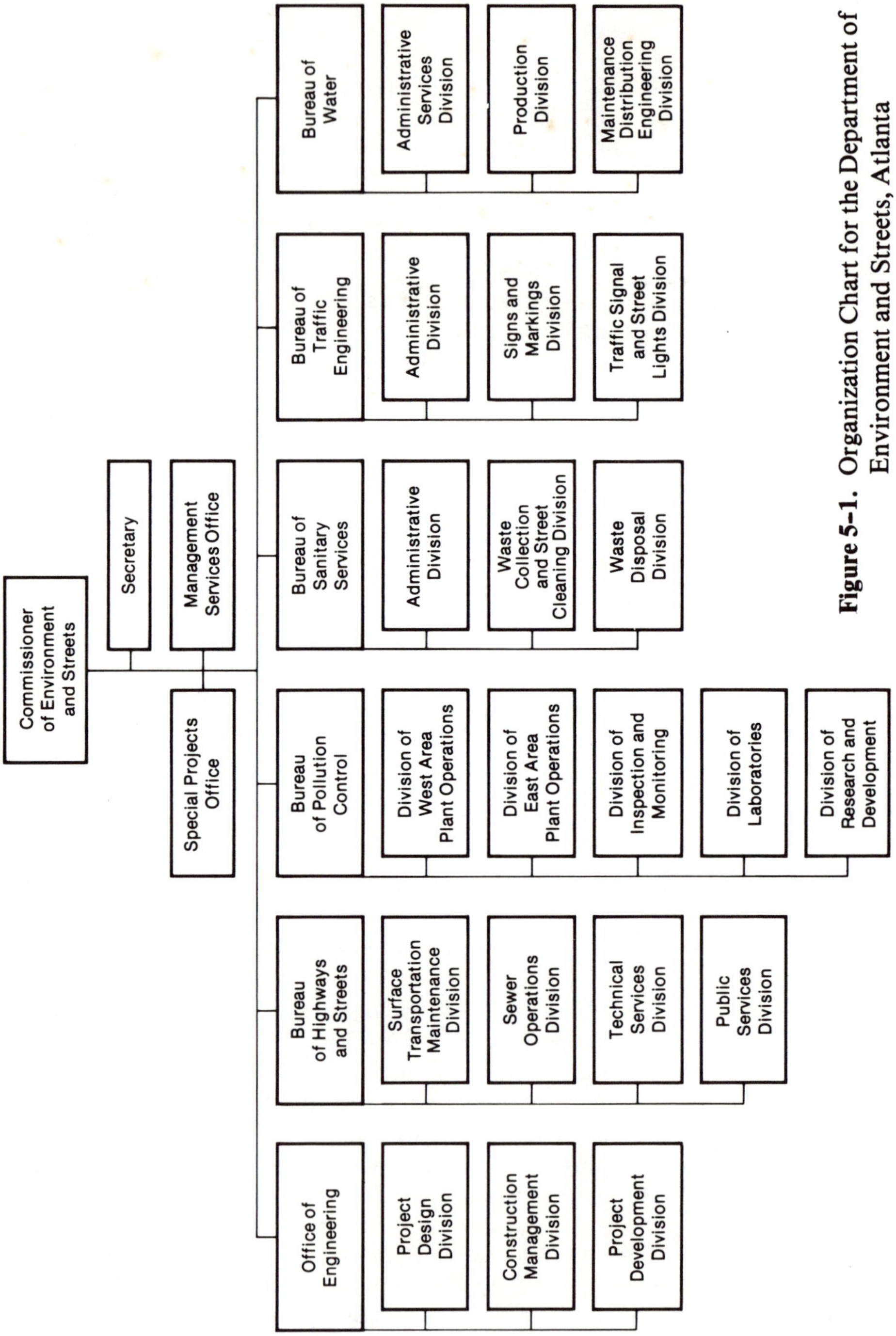

Figure 5-1. Organization Chart for the Department of Environment and Streets, Atlanta

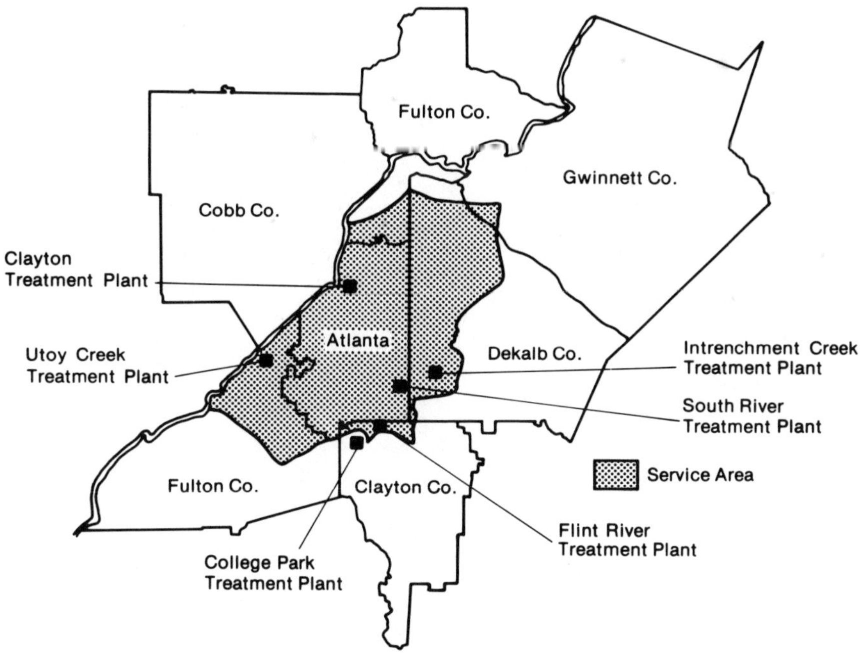

Figure 5-2. Map of Atlanta Metropolitan Sewer System Service Area

sewer-charge and surcharge billings, which are the main revenue source for the Bureau of Pollution Control, are based on water-meter readings and are sent out with the water bills. Although sewer bills are based on water usage, an allowance is made for firms whose wastewater flow is significantly less than their water-meter reading because of water use in their products. Procedures for calculating sewer-charge and surcharge levies are described later in this chapter. Both bureaus are supposed to maintain financial self-sufficiency, but, according to bureau representatives, water-supply revenues are used to some extent to subsidize pollution-control expenditures. The capital costs of upgrading and altering the treatment facilities are subsidized by federal and state funds to the extent possible.[1] The local share of these costs is raised through floating revenue bonds, with the bond payments being met from the combined water and sewer-service revenues.

Table 5-1 presents actual and estimated water and sewer-system revenues and expenditures for 1975-1980. This table shows revenues in excess of the amount required to cover operating expenses and the costs of servicing revenue bonds. The surplus is transferred to the city to service general-obligation bonds.

Table 5-1
Water and Sewer System Revenues and Expenditures in Atlanta
(*thousands of dollars*)

	Actual			*Estimated*		
	1975	*1976*	*1977*	*1978*	*1979*	*1980*
Revenues:						
Water sales	17,563	20,630	25,909	26,877	35,158	36,012
Sewer contracts[a]	2,795	2,999	2,833	3,448	3,511	3,866
Sewer-service charges	10,516	11,193	12,324	12,677	23,776	24,354
Other	429	359	488	240	380	331
Total revenues	31,303	35,180	41,554	43,242	62,825	64,563
Operating expenses:						
Water operations	10,829	12,233	13,295	16,800	18,003	20,485
Sewer operations	6,374	7,126	7,661	10,040	11,275	12,700
Total operating expenses	17,203	19,359	20,956	26,840	29,278	33,185
Debt service on existing lien bonds	2,128	1,829	1,858	1,963	1,419	1,227
Revenues available for debt service on water and sewerage revenue bonds	11,972	13,992	18,740	14,439	32,128	30,151
Maximum annual debt service on outstanding revenue bonds and series 1978 revenue bonds[b]	—	—	9,176	9,176	11,359	11,359

Source: City of Atlanta, Georgia, Water and Sewerage Revenue Bond, Series 1978, prospectus issued December 1978.

[a] Sewer contracts with other counties.

[b] These figures are based on the assumption that 7 percent interest will be paid on series 1978 bonds. They do not include the debt service on the $1.9 million general-obligation water bonds. When the water and sewer system has excess revenues, the city transfers money from the Water and Sewerage Revenue Fund to the General Obligation Debt Service Fund. In 1977, $242,437 was so transferred.

Wastewater Management

Demand

The Bureau of Pollution Control provides wastewater treatment services to residential, commercial, and industrial users. Roughly 85 percent of the total flow comes from residential and commercial sources; the remaining 15 percent is accounted for by industrial dischargers. Approximately 100,000 households receive waste-treatment services in the service area, with 91,000 of these households located in the city of Atlanta.

There are approximately 200 industrial users in the system. All are monitored, to ensure that prohibited substances are not discharged and to test whether normal concentrations (250 ppm for both BOD and SS) are exceeded. About 67 percent of these industrial users do not exceed these concentrations and thus pay a surcharge. At this time, the only NPDES permit issued to a plant in Atlanta is that held by Delta Airlines for an aircraft-maintenance operation.

In 1977, the wastewater received by Atlanta treatment plants had an average BOD level of 172 ppm and an average SS level of 184 ppm. Total daily flow is about 130 million gallons.

The city of Atlanta obtains its water from the Chattahoochie River at points adjacent to the city limits. River flow is regulated at the Buford Dam, located thirty miles above the city's intake facilities. The uniformity of river flow is also maintained by a control dam, seventeen miles above the city's intake. The city has two water-treatment plants with a combined capacity of 132 million gallons per day.

Supply Characteristics

The city of Atlanta has entered into three types of treatment arrangements: treating sewage from within the city limits; treating other municipalities' sewage; and contracting for treatment by other municipalities.

Through 1974, the city of Atlanta and Fulton County had a thirty-year contract that was the basis of joint ventures in sewerage. Upon the termination of this contract, the city moved toward separate contractual agreements with each of the municipalities. The typical arrangement is for a municipality to pay a proportionate share of both a plant's capital costs and its operating costs. The new contracts take into account the quality of the sewage for the first time.

While Atlanta plants treat sewage from Fulton County, DeKalb County, and Forest Park, sewage from Atlanta flows to two treatment plants outside the city limits. An average 0.164 million gallons per day (mgd) flows to Cobb County, which charges the city \$.45 per thousand gallons. At the Camp Creek Plant in Fulton County, 23 percent of the flow of about 1.67 mgd comes from Atlanta. Atlanta has contributed to the initial capital costs of Camp Creek and is charged a percentage of monthly O&M costs.

Atlanta's own collection and treatment facilities consist of approximately two thousand miles of sewers and five wastewater-treatment plants. Total capacity reserved for the city of Atlanta in its own plants is approximately 144 mgd, with an additional capacity of 7 mgd reserved for the city at Fulton County's Camp Creek Plant. Daily flow at all five of the city's treatment plants is well below design specifications. Atlanta's treatment plants and the sources of their flow are described in table 5-2. Statistical data for these five treatment plants for 1977 are presented in table 5-3, and data on their operating and maintenance costs from 1969 to 1977 are presented in table 5-4. Prior to 1974, the city maintained several other small treatment facilities, with a total daily flow of a little over 3 percent of total daily flow in 1969.

Table 5-2
Atlanta Treatment Plants

R.M. Clayton

Constructed in 1938 as a 42 mgd primary-treatment plant, the plant was modified in 1977 to a 120 mgd activated-sludge facility.

Percent Flow Contribution: Atlanta, 54%; Fulton County, 4%; DeKalb County, 42%

South River

Constructed in 1936 as a 5.5 mgd trickling-filter plant, the plant underwent major modification in 1962 to increase capacity to 17.5 mgd.

Percent Flow Contribution: Atlanta, 81%; Fulton County, 18%; Forest Park, 1%

Intrenchment Creek

The original plant was an Imhoff-tank system constructed in 1910 (capacity unknown). Major modifications took place in 1963, converting the plant to fixed-nozzle trickling filter (capacity unknown). The second modification took place in 1963, converting the plant to a high-rate trickling-filter system with a capacity of 20 mgd.

Percent Flow Contribution: Atlanta, 79%; DeKalb County, 21%

Flint River

Constructed in 1944 as a 2 mgd trickling-filter facility. Relocated in 1971, the plant now provides a 6 mgd activated-sludge facility.

Percent Flow Contribution: Atlanta, 28%; Fulton County, 72%

Utoy Creek

Constructed in 1937 as a 2 mgd primary facility, it underwent modification in 1956 to bring the capacity to 9 mgd. The most recent modifications were completed in 1973, bringing the capacity to 30 mgd activated-sludge treatment.

Percent Flow Contribution: Atlanta, 60%; Fulton County, 40%.

Table 5-3
Statistical Data on Atlanta's Treatment Plants, 1977

	Flow		BOD				TSS				Cost[a]		
Plant	*Total (mgd)*	*Treated (mgd)*	*Influent (mg/l)*	*Effluent (mg/l)*	*Removed (kg/day)*	*Discharge to Stream (kg/day)*	*Influent (mg/l)*	*Effluent (mg/l)*	*Removed (kg/day)*	*Discharge to Stream (kg/day)*	*$/mg[d] Treated*	*$/kg BOD or TSS Removed[e]*	*Total O&M ($)*
R.M. Clayton	86.05	85.85	166	30	44,166	10,206	194	83	36,611	27,491	111.82	0.1[illegible]9	3,503,885
Flint River	2.43	2.38	206	28	1,613	284	143	33	992	315	530.27	0.434	460,645
Intrenchment Creek	13.37	12.60	190	33	7,507	2,245	152	42	4,952	2,787	139.64	0.1[illegible]1	642,217
South River	14.00	13.99	196	27	8,677	1,678	169	53	6,014	2,873	203.94	0.1[illegible]4	1,041,323
Utoy Creek	12.77	12.77	163	24	5,694	2,080	168	44	5,989	2,103	226.84	0.2[illegible]8	1,057,324
Hollywood Road and Pump Stations[b]	1.42	1.42	126	39	487	196	131	60	407	327	722.30	1.1[illegible]4	312,758
Total or Average[c]	130.04	129.01	172	34	68,144	16,689	184	73	54,965	35,896	149.04	0.1[illegible]7	7,018,152

Source: Atlanta Bureau of Pollution Control.

[a] These figures include operation and maintenance expenditures plus overhead. No capital costs are included.

[b] Cost figures for Hollywood Road Package Plant include operation and maintenance cost for all pump stations within city limits; therefore, figures do not depict an accurate cost for treatment.

[c] Averages were weighted according to total flow.

[d] $/mg was calculated by dividing *total* O&M by million gallons of flow treated.

[e] $/kg BOD or TSS removed were calculated as follows: (total $O&M ÷ 365 days) ÷ (BOD kg/day + TSS kg/day). They thus represent the combined cost of removing BOD and TSS if *no* O&M costs are attributed to flow.

Table 5-4
Cost of Operating and Maintenance
(dollars)

Plant	*1969*	*1970*	*1971*	*1972*	*1973*	*1974*	*1975*	*1976*	*1977*
R.M. Clayton	230,648	293,440	329,235	552,414	910,056	1,443,545	2,185,345	2,442,707	3,503,885
Flint River	69,592	66,025	120,514	257,448	299,965	341,839	370,155	378,068	460,645
Intrenchment Creek	311,790	330,767	359,449	389,047	409,456	458,870	548,773	506,858	642,217
South River	537,014	651,600	681,416	789,361	771,457	804,726	958,591	920,053	1,041,323
Utoy Creek	105,987	120,176	126,216	152,277	214,882	671,202	838,948	913,621	1,057,324
Five-plant total	1,255,031	1,462,008	1,616,830	2,140,547	2,605,816	3,720,182	4,901,812	5,161,307	6,705,394

As is evident in table 5-4, costs of operating the five treatment plants have increased dramatically, while flows have increased only moderately, from 113.69 mgd to 128.62 mgd from 1969 to 1977. The Clayton Plant has the lowest cost of treatment in all cost categories, whereas Flint River is the most expensive. It should be pointed out, however, that table 5-3's figures on dollars per million gallons treated and dollars per kilogram of BOD and TSS removed are not true statistics since they do not segregate costs attributable to flow from those attributable to removal but attribute *all* O&M costs to one or the other. The figures are used to facilitate cost comparisons between plants for a given year, as well as cost comparisons for a given plant over time.

In 1972 and 1973, the Environmental Protection Division (EPD) in Georgia's Department of Natural Resources directed the city of Atlanta either to discontinue the discharge of wastewater into Intrenchment Creek, South River, and Flint River or to improve the degree of treatment at the three plants discharging into these waters. In order to meet new water-quality standards, the EDP also mandated the retention and treatment of occasional sewer-system overflows occurring during periods of heavy rainfall. On the basis of a 1974 report prepared by Black, Crow and Eidness, Inc., Engineers, and a 1976 report by the city's Bureau of Pollution Control, the city has embarked on the Three Rivers Project, which will do three things:

1. Wastewater treated at the Intrenchment Creek, South River, and Flint River Treatment Plants will be diverted to the Chattahoochie River (the source of all the city's water).
2. Treatment at these three plants will be improved.
3. Initial or "first flush" sewer-system overflows will be treated before discharge.

According to construction schedules, the project is to be completed by 1982. Although original cost estimates were $80.8 million, the most recent estimates are $208 million. Of this, EPA grants will cover $155 million, the city of Atlanta will pay $40 million, and the remaining $13 million will be allocated among other localities.

The bureau has had a number of arguments with the state's Environmental Protection Division (the agency responsible for permit administration) over operations of the Clayton Treatment Plant. Clayton, although recently upgraded with EPA Construction Grants Program funds, has been subject to continual breakdowns. The plant breakdowns have been attributed to design problems, to difficulty in obtaining equipment during the upgrading project, and to O&M problems resulting from inadequate training of treatment-plant operators. The bureau complains about the loss of capable workers to industry but contends that the breakdowns are diminishing as some of the earlier problems are remedied.

Changes in Atlanta industry's treatment methods and treatment levels have resulted not so much from the industrial surcharge program as from the more stringent permit requirements imposed by Georgia's Environmental Protection Division. These requirements are imposed only on the city's treatment plants, not on industries discharging their waste into the city's sewers. However, because of the state requirements, the city has issued an ordinance regarding industrial discharges. These discharges are required to be neutralized to a pH level of 5.5 to 9.5, but some industries have difficulty meeting this standard. The phenol limit in the ordinance is stated at 0.5 parts per million, but wood-preserving operations, photoprocessing, and chemical companies in the Utoy Basin sometimes exceed this limit. Platers frequently discharge heavy metals, but the ordinance does not cover heavy metals at this time.

Industrial Surcharge Program

History

Although planning for Atlanta's industrial surcharge program began in 1965, it was not put into effect until 1967. The purpose of the surcharge program was twofold: to ensure that the Bureau of Pollution Control collected adequate revenues for the treatment of industrial wastes and to ensure equity between residential and industrial users. The city council also expressed a concern with the viability of small businesses and wanted to ensure that the charge levels adopted did not result in bankruptcy for small firms.

Prior to the adoption of the surcharge scheme, the bureau sampled all industries discharging wastes into the public system. This monitoring effort enabled the bureau to identify all plants with effluents above normal strength. The need for adequate surcharge revenues to cover costs and the extent of the burden for individual dischargers were taken into consideration in devising the surcharge formula.

The provisions of the charge scheme, the surcharge formula, and the charge levels were adopted as a city ordinance by the city council and the mayor during the course of the 1967 annual budget hearings. Once the ordinance was passed, the bureau undertook a publicity campaign to acquaint industrial users with its terms. Articles describing the program were published in the newspapers, and all industrial dischargers were visited individually to explain the purposes and the terms of the surcharge program.

Table 5-5
Atlanta Water Rates, 1978

Minimum bill, 400 cubic feet	$3.10		($.7750 per hundred cubic feet)
Next 1,600 cubic feet	$3.10	+	$1.17 per hundred cubic feet over 400
Next 5,000 cubic feet	$21.82	+	$1.18 per hundred cubic feet over 2,000
Next 10,000 cubic feet	$80.82	+	$0.78 per hundred cubic feet over 7,000
Next 50,000 cubic feet	$158.82	+	$0.64 per hundred cubic feet over 17,000
Next 33,000 cubic feet	$478.82	+	$0.52 per hundred cubic feet over 67,000
Next 100,000 cubic feet	$650.42	+	$0.48 per hundred cubic feet over 100,000

Provisions

Charges for water supply are calculated on a declining block basis, with significantly lower rates for quantities above specified levels. The current rate schedule is presented in table 5-5.

Sewer charges were calculated at 50 percent of the water bill until 1971, when a flat charge of thirty-five cents per hundred cubic feet was levied. This was raised to forty-four cents per hundred cubic feet in 1976, and recently raised again to eighty-eight cents per hundred cubic feet. Sewer charges have always taken into account capital as well as operating costs.

All plants with discharges in excess of normal concentrations are subject to surcharge levies. There are approximately sixty-five such plants participating in the program. The surcharges are calculated according to the following formula:

$$\$/\text{ccf} = (B + C) \times (0.006238) \times \text{cost factor}$$

where B = TSS − 250 (in parts per million); C = BOD − 250 (in parts per million); and 0.006238 is the water weight factor. The base rate for wastewater flow includes contributions for handling BOD and TSS at 250 mg/l.

The cost factors are based on actual costs from the previous year; they represent the average cost of removing a pound of BOD or a pound of TSS in the entire sewer system. It is assumed that BOD and SS removal costs are equal. Initially only O&M costs were factored into the calculations for the surcharge, but, beginning in 1978, capital costs were also included in the surcharge as well as the basic sewer-charge rates. It should be noted that capital costs cover treatment facilities only; the costs associated with sewers and pumping stations are handled by another department with funds from general revenue. Table 5-6 presents surcharge cost factors as well as sewer

Table 5-6
Atlanta Surcharge Cost Factors and Sewer-Charge and Surcharge Revenues
(dollars)

Year	*Surcharge Cost Factor*	*Sewer-Charge and Surcharge Revenues*
1967	0.013	3,651,239
1968	0.013	9,421,291
1969	0.020	9,763,286
1970	0.025	10,322,861
1971	0.025	10,662,995
1972	0.035	12,634,281
1973	0.035	12,586,319
1974	0.053	12,366,425
1975	0.053	10,516,000
1976	0.053	11,193,000
1977	0.081	12,324,000
1978	0.105	12,677,000

charge and surcharge receipts for the period 1967–1978. Over the period, the cost factor increased from approximately one cent per pound to over ten cents per pound.

The city council must approve any increases in the cost factor, but the bureau itself determines when those rates must be raised. According to their municipal bond provisions, their revenues must be at least one and a third times greater than the principal and interest paid on all outstanding bonds plus bonds about to be issued. If it appears that the revenues will not meet this requirement, the bureau writes a new ordinance, which then goes before the city council. Approval is generally pro forma.

A surcharge data sheet, used by the bureau for calculating the surcharge rates for individual dischargers, is presented in figure 5-3. The surcharge calculations are performed by the Bureau of Water Supply and are based on the surcharge rates and metered water use. Water use is adjusted to reflect the quantities of water retained in the product, based on data collected at the plants. Water bills, sewer-charge bills, and surcharge bills are submitted to industrial users on a monthly basis by the Bureau of Water Supply. It is the responsibility of the Bureau of Pollution Control to advise the Bureau of Water Supply of changes in the surcharge rates caused by monitored changes in effluent concentrations or changes in the cost factors.

Monitoring

The Division of Inspection and Monitoring is responsible for sampling industrial effluents, and the analyses are performed in laboratories maintained by the Department of Environment and Streets. Sampling is supposed to occur twice a year, without advance warning, to prevent industries from changing their production processes during the sampling periods. Sampling frequency decreases, however, when the division is particularly busy with other projects, such as the Three Rivers Project.

If industries disagree with the results of the monitoring effort, they may request another sample at their own expense. Resampling costs range from $300 to $500. Resampling is done with advance notice, and a plant representative is present during the sampling period. If the results of the resample

INDUSTRY__________

ADDRESS__________

DESCRIPTION__________

DATE__________

TYPE TEST		P.P.M.
TSS	– 250*	
BOD	– 250*	
TOTAL		

COMMENTS:
*REDUCTION FOR NORMAL SANITARY WASTE
________% OF TOTAL WATER CONSUMPTION RETURNED TO SEWERAGE SYSTEM

X for 19_____ = $________
Y for 19_____ = $________

B = THE EXCESS OF 250 PPM OF TSS
C = THE EXCESS OF 250 PPM OF BOD
FACTOR X = THE ANNUAL AVERAGE COST OF REMOVAL OF ONE POUND OF TSS
FACTOR Y = THE ANNUAL AVERAGE COST OF REMOVAL OF ONE POUND OF BOD

	WT. PER 100 CU.FT. WATER 6,238 ÷ 1,000,000		COST FACTOR		COST PER 100 CU. FT. OF WATER CONSUMPTION RETURNED TO THE SEWERAGE SYSTEM
( ) B + C	× (.006238)	×	( ) XY	=	$ ________

COMPUTED BY__________

CHECKED BY __________

____
SUPERVISOR OF INDUSTRIAL WASTE AND SEWER SERVICE CHARGE

Figure 5-3 City of Atlanta, Bureau of Water Pollution Control, Surcharge Data Sheet

indicate lower concentrations than in the initial sample, the surcharge rate is lowered accordingly. The city ordinance establishing the industrial surcharge program requires the division to comply with any requests for resampling.

Division representatives feel that requests for resampling are made lightly, and that resampling activities interfere with the performance of other duties. To circumvent this problem, a program of self-monitoring has been proposed. According to this proposal, firms are to fill out questionnaire that describes the characteristics of their operations and the results of their effluent analysis. The division will perform spot checks to verify the validity of the information reported and will then use the questionnaire information as the basis for calculating surcharge rates. The questionnaire was adopted by revision of the city ordinance in 1978.

Enforcement has not presented a major problem. The ordinance does provide for the termination of service at the discretion of the Division. Although shutoff has been used effectively as a threat, it has in fact occurred only once since the beginning of the surcharge program.[2] Nonpayment occurs only when the monitoring results are in dispute, and accounts are normally settled once the resampling has taken place.

Impact on the Agency

The major effect of the surcharge program has been an increase in the revenues of the bureau. Direct program-operating costs (excluding laboratory and treatment costs) are under $100,000 per year, while surcharge revenues are several million dollars.

Impact on Industry

The major effect of the surcharge program on the participating firms has been an increase in the costs of receiving sewerage collection and treatment services. According to bureau representatives, firms have improved their housekeeping in response to higher costs, but only on a sporadic basis. The bureau believes that industries have not significantly reduced their flows or their effluent concentrations, although the statistics in table 5-7 indicate that there has been some drop in BOD and TSS levels. This may be explained, however, by an increasing residential population and wastewater discharges. Pretreatment has been considered by some plants, according to division representatives, but this option has been rejected because the charge bills are lower than the costs of self-treatment. According to the same source, product and process changes have not resulted from the pro-

Table 5–7
Average BOD and TSS Levels in Influent Coming to Atlanta's Treatment Plants

Year	*BOD(ppm)*	*TSS (ppm)*
1969	204	NA
1970	253	NA
1971	245	NA
1972	249	217
1973	235	196
1974	225	172
1975	148	144
1976	186	152
1977	172	184

gram. Pretreatment is expected to become more prevalent in the future as a result of federal regulations, but not as a result of the surcharge program.

Firms have not elected to leave the system. According to division representatives, no firms have relocated or gone out of business because of the surcharge. Since the imposition of the program, the number of firms using the city's wastewater-treatment services has remained the same. Only one firm has its own NPDES permit.

Summary

Atlanta has a straightforward industrial surcharge program, which has been in effect since 1967. Charges to industry for the city's sewerage collection and treatment services are based on both volume and strength. Approximately sixty-five firms participate in the program. Another 135 discharge into public sewers, and are monitored for compliance with effluent limitations stated by the city ordinance, but do not have discharges in excess of normal and therefore do not pay surcharges. Sampling and monitoring are performed by the city. According to city officials, the program has been well accepted by the industrial participants, and enforcement has not been a problem.

The major goal of the program has been the generation of adequate revenues to ensure financial self-sufficiency. This goal has been met. According to the system representatives, reductions in waste volumes and effluent concentrations have not occurred to any significant degree, and effects on the participating industries have been minor.

Notes

1. For example, 75 percent of the construction costs for the Three Rivers project will be covered by EPA grants, and the remaining 25 percent will be paid by the various localities benefiting from the project.

2. The shutoff occurred with a construction operation that was washing out cement trucks.

6 Chicago

Agency Overview

The Metropolitan Sanitary District of Greater Chicago (the district) was founded in 1886 by the Illinois State Legislature. This action was prompted by an epidemic that was directly attributable to contamination of the city's water supply by sewage flowing via the Chicago and Calumet Rivers into Lake Michigan after a heavy rain. Approximately 12 percent of the city's population died from waterborne diseases such as cholera, typhoid, and dysentery. As a result, the founders of the district devised a plan to reverse the flows of the Chicago and Calumet Rivers away from Lake Michigan. A series of canals was constructed, diverting flows to the Mississippi River via the Des Plaines and Illinois Rivers. The large volume of flow was considered sufficient to maintain water quality despite the fact that all wastewater generated in the city was discharged directly into the canal system.

Although these flow reversals prevented contamination of Lake Michigan, protecting the city's water supply, there was concern that the water level in the lake would be lowered. As the result of suits brought by neighboring states, the flow through the canals was limited and controlling locks were installed. These flow reductions had an adverse impact on the water quality in the canal system, and plans for wastewater treatment began shortly thereafter. The first plants were built about 1930 and provided primary and secondary treatment.

In addition to providing wastewater management, the district operates the canal system, providing flood control and control over river navigation.

The service area of the district has expanded over the years so that it now encompasses approximately 91 percent of the area of Cook County, an area of approximately 866 square miles. A total of 125 communities, including the city of Chicago, are served by the district. The domestic load on the system is approximately 5.4 million persons, with an additional nondomestic load equivalent to 4.5 million persons.

The district is an independent governmental agency chartered by the state to provide services to a number of local jurisdictions within Cook County.

A board of nine elected commissioners establishes the policies and goals of the district. Three of the commissioners are elected at large every two years for a six-year term. The board elects from its members a president, a vice-president, and chairman of finance. The chief executive officer under the board is the general superintendent, who manages and controls all affairs and property of the district. The general superintendent appoints all department heads, with the consent of the board. The organization chart in figure 6–1 shows the various departments within the district.

The district is responsible for the operation and maintenance of only its own wastewater-management facilities. There is no direct district control over the local sewers that connect to its interceptors at more than 10,000 points. However, all new local sewers that feed into the system must meet minimum standards set by the district.

The Research and Development Department is responsible for implementing the industrial surcharge program, which is the subject of this study. The department also has responsibility for developing the user-charge and industrial cost-recovery systems for the district. The district has direct authority over all industries discharging into the system whether the discharge is made directly into a district interceptor or into a local system.

The district is supported by direct taxation of users through the ad valorem real and personal property tax. Cook County is responsible for collecting these taxes and distributing them respectively among all the taxing bodies within the county.

The county assessor determines assessed value based on a percentage of market value once every four years. The district submits its tax levy requirements to the county, which then sets the tax rate required to generate sufficient income. The taxes are collected by the county and transmitted to the district. This system means that users in all 125 member communities are taxed for the district's services on the basis of their assessed value rather than their usage of water and sewer services. Only the supplemented revenues from the industrial surcharge program (described later) are based on usage.

The district's budget is developed by the general superintendent and reviewed and approved by the board. The board, in turn, is directly accountable to the electorate.

The district may also issue bonds to raise funds for capital investments. State-legislature approval, or even a statewide referendum, may be required, depending on the size of a proposed bond issue. The district is currently selling bonds under a 1973 authorization by a statewide referendum to sell $384 million worth of bonds. The district's bonds at present carry a triple A rating.

The District's finances are handled through five funds:

1. The Capital Projects Fund, which receives revenues from grants and the sale of bonds, covers capital expenditures. In 1977 this fund had revenues of approximately $175 million.

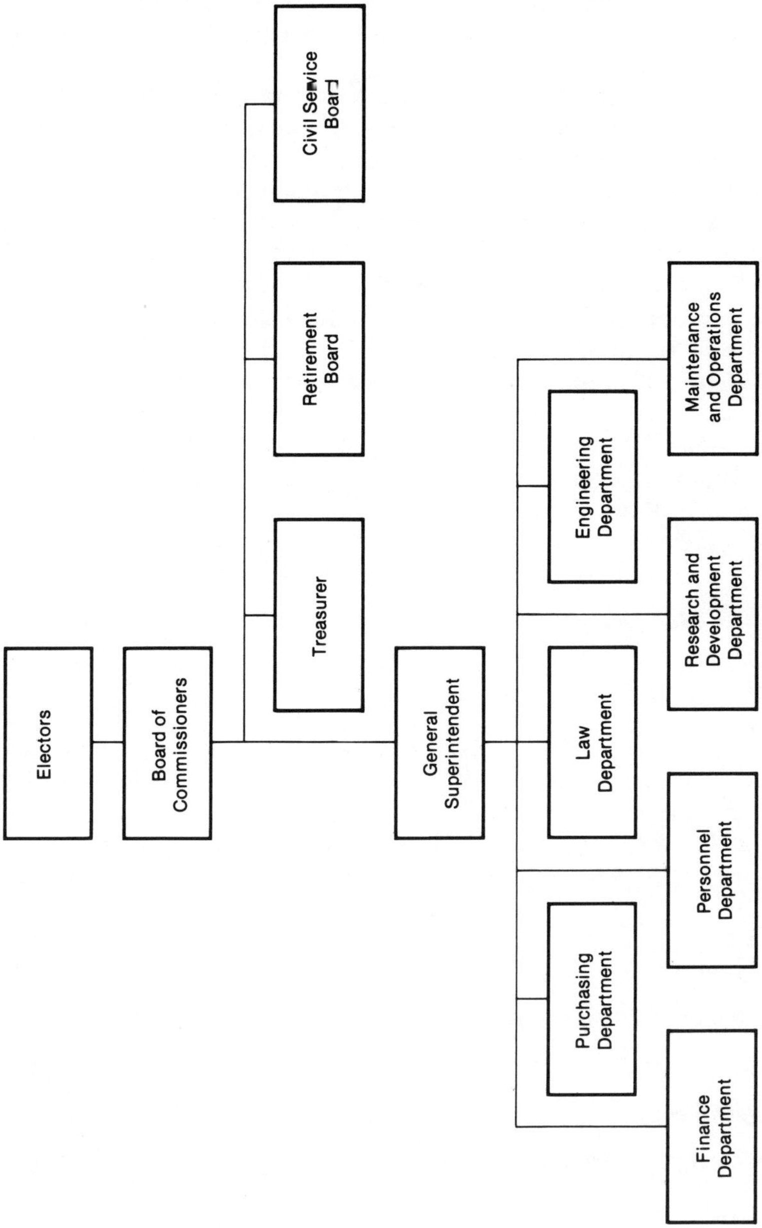

Figure 6–1. Organization Chart for the Metropolitan Sanitary District of Greater Chicago

2. The General Corporate Fund, which receives revenues from the property tax, the industrial surcharge, land rentals and fertilizer sales, and operating and maintenance and administrative costs, had 1977 revenues of approximately $85 million.
3. The Debt Service Fund, which receives revenues primarily from the property tax, covers redemption and interest on bonds. Its 1977 revenues were approximately $38 million.
4. The Special Revenues Fund, which receives revenues primarily from the property tax, covers both the pension fund and claims against the district for such things as damage caused by backed-up sewers. The Special Revenues Fund received $69 million in revenues in 1977.
5. The Trust and Agency Fund accounts for resources received and held by the district as an agent or trustee. This fund had a balance at the end of 1977 of $35 million and no outlays during that year.

The industrial surcharge revenues of approximately $5.6 million account for about 7 percent of the total revenues collected for operation of the district's facilities.

Wastewater Management

Demand

The total average daily flow from the district's wastewater-management facilities is approximately 1.5 billion gallons per day. In 1978 the district conducted a usage study to be used in generating a user-charge system to comply with federal regulations. This unpublished study estimated that the various user groups are responsible for the following percentages of the flow:

Residential	41 percent
Small industrial and commercial	15 percent
Large industrial and commercial	4 percent
Governmental	2 percent
Tax-exempt (churches, schools, and the like)	2 percent
Unattributed	36 percent

The unattributed portion of the flow represents excessive infiltration inflow and storm-water flows. Under the proposed user-charge systems, each user group will pay a proportionate share of the costs of handling these unattributed flows.

The district has a very strong sewage and waste-control ordinance, which specifies minimum quality standards for wastes discharged into any sewer leading to district facilities. This ordinance is actively enforced, greatly reducing the potential for shock loads at the treatment plants.

Supply Characteristics

The district boasts the largest sewage-treatment facility in the world. The Southwest Plant, which handles the wastes from the central part of the district by means of conventional activated-sludge secondary treatment, has an average flow of 800 to 900 mgd, with a wet-weather capacity well in excess of a billion gallons per day. The Southwest Plant continually meets or exceeds the water-quality standards, specified in the NPDES permit, of 12 ppm suspended solids and 10 ppm BOD, eliminating the need for advanced treatment processes.

The Southwest Plant is the largest of the district's seven treatment plants. These plants, in combination with twenty-six pump stations and nearly 500 miles of interceptor sewers, make up the district's sewerage facilities (figure 6-2). The total average wastewater flow handled at the district's facilities is almost 1.5 billion gallons per day, with a daily removal of 1,000 metric tons of solids.

Performance at all district treatment facilities is impressive. The plants are well maintained and very carefully operated. Sufficient funding has been allocated to rehabilitate or replace equipment before it fails, so there are almost no mechanical failures that result in discharges of inadequately treated wastes.

Combined sewer overflow, which occurs almost weekly, does constitute a serious pollution problem. The district, which is also responsible for flood control within its service area, is currently in the process of constructing a massive deep tunnel and storage reservoir system to all but eliminate the problem of combined sewer overflows.

The district also operates the 71 miles of navigable waterways that make up the river and canal network devised in 1886. The district is responsible for the maintenance of water quality within these waterways. As a result of extensive study, it has been concluded that, in some portions of the system, instream aeration will be the best method for attaining minimum dissolved oxygen standards for waterways, as established by the Illinois Pollution Control Board. In other parts of the district, advanced wastewater-treatment facilities are being installed.

In 1977, the district General Corporate Fund had total expenditures of

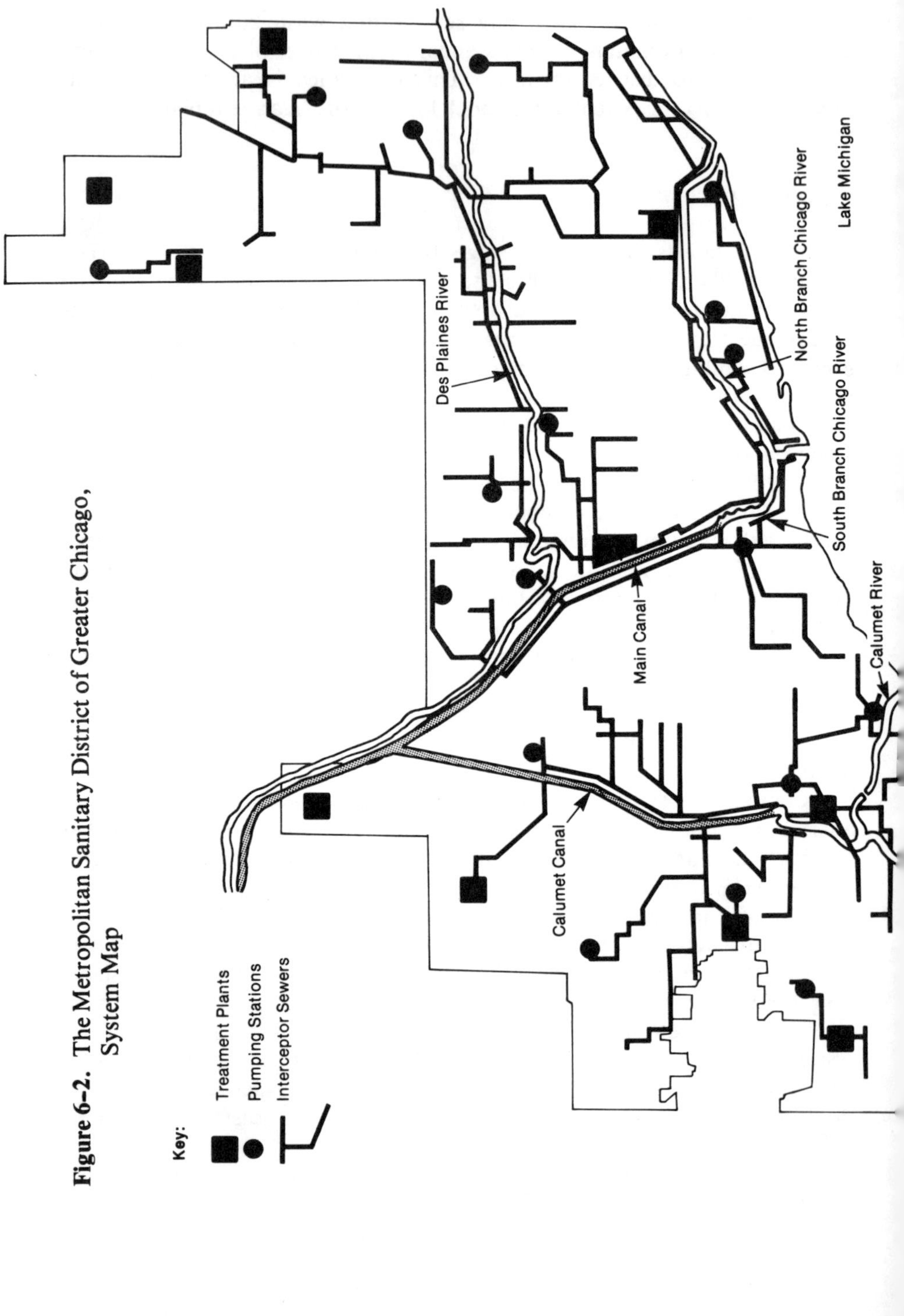

Figure 6–2. The Metropolitan Sanitary District of Greater Chicago, System Map

$82 million, of which slightly more than $62 million were for the operation and maintenance of facilities. The remaining $20 million was spent on administration.

Industrial Surcharge Program

History

The industrial surcharge program was initiated in 1969 and became effective 31 December 1970. The goal of the surcharge program has been to have industry pay a more equitable share of the costs of wastewater treatment and to reduce the burden on residential taxpayers. The surcharge program was never intended to serve as an incentive for industries to reduce wastewater flows or loadings.

Provisions

Industrial surcharges are levied on plants whose property-tax payments accruing to the district are less than the estimated cost of treating the plant's industrial waste. Estimates of the costs of treating a plant's wastewater are calculated from unit costs developed annually by the district's staff on the basis of average costs for the district as a whole. In 1978 the unit prices used in computing the liability of individual plants were as follows:

$0.042 per 1,000 gallons of flow

$0.041 per pound of BOD

$0.056 per pound of suspended solids

Firms that combine industrial and domestic waste streams were given a deductible allowance of $0.008 per employee per day worked.

The unit prices calculated by the district cover both capital and O&M costs. The unit price set for handling and removing suspended solids is based on actual costs. In contrast, the unit price for BOD loading is based on the theoretical capacity and air requirements needed to treat a given waste. The unit price levied on flow is based on both theoretical and actual costs: the theoretical cost for providing capacity at plants, the actual cost of providing capacity in the sewers, and actual O&M costs.

The surcharge program currently affects approximately 2,000 manufacturing concerns in the Greater Chicago area. Under the current regulations, commercial and nonmanufacturing industries are not subject to the charge, but such firms will be included once the new user-charge system is adopted next summer. The number of firms subject to the charge is expected to increase to 6,000 when the district revises its user-charge system to comply with the requirements of the 1977 Clean Water Act. The user charge, however, will still be collected by means of an ad valorem tax, as it is now, and most of the firms added under the new system are not likely to pay a surcharge, as their property-tax payments accruing to the district will, in most cases, cover the costs of treating their wastewaters.

Of the 2,000 firms currently subject to the surcharges, most do not have to pay the charge. The payers consist of a group of approximately 350 firms who are required to submit a surcharge calculation and payment to the district monthly. These 350 firms tend to be food processors, electroplaters, and a wide range of chemical plants. Their flow can be determined either by direct measurement or by meter reading, their BOD and suspended solids levels are determined by samples obtained once a year by the firms themselves.

The nonpaying or exempt firms are divided into two types, Class B and Class C. Class B exempt firms are those that do not discharge process water into the district's sewers. Such plants must certify that they discharge only domestic waste or cooling water. They must file a form every year, but they are not required to sample their wastewater.

Class C exempt firms are those that discharge industrial waste into the district's system but are not liable to surcharge payments because the taxes paid are sufficient to cover treatment costs. Firms in this category are required to file a form annually, which requires that the firm conduct a full program for seven consecutive twenty-four-hour days. Class C exempt firms are heavy manufacturers, many of whom are subject to toxic regulations requiring pretreatment of their wastes.

Monitoring

The surcharge program is based on a self-monitoring system. All payers and Class C exempt firms must have BOD and suspended solids analyses conducted on proportional samples collected hourly for seven consecutive days. Thus, discharges must be monitored one complete week per year. Some firms collect and test samples every month because of great variations in their wastewater, but all that is required is annual sampling. Each plant schedules its own sampling program. The plants generally hire consultants to collect the samples, test them, and fill out the required forms.

The district employs three teams strictly for surcharge verification. The teams make unannounced visits to large payers approximately once a year. The schedules for visits are varied to reduce the firms' ability to modify process during the sampling period. Other payers may be visited every two or three years. Large nonpayers may be visited once in a lifetime, but the district does not generally check on small nonpayers.

During these visits, the district will split a sample with the plant but will not allow the plant to use the results of the district's analyses in filing under the surcharge. Thus, although the district pays the collection costs for these samples, all firms must at least bear the costs of having the samples analyzed. The policy of requiring double testing of the samples collected for monitoring gives the firms an opportunity to check on the accuracy of the district's analyses. If there is a large discrepancy between the results achieved by the district and those achieved by the firm or its consultant, a negotiation is arranged, and, in some cases, a repeat visit by the sampling team may occur. The district is aware that firms frequently attempt to minimize waste production when the verification team is monitoring the plant. However, extreme variations or inconsistencies resulting from drastic process changes can often be detected because of inspectors' knowledge of data from previous years or from similar plants, because of information from informants, or because of inspectors' observation of plant processes. When the surcharge program began, disagreements between the firms and the district occurred more often than they do now.

The district's legal department is responsible for the enforcement of the industrial waste discharge program. When the program was implemented, several firms challenged the ordinance. One large payer took the challenge all the way to the Illinois Supreme Court. Once the case was decided in the district's favor, the firm was required to pay several million dollars in back payments, penalties, and interest. Opposition to the program decreased dramatically.

In at least one instance, a firm that was under order to file as a payer agreed to do so only after assurance that its competitors would be charged also.

Impact on the Agency

The district collects between $5 million and $6 million annually from the surcharge program. Although exact figures were not known, the total administrative cost of the program was estimated at 15 percent to 20 percent of the revenues.

The district's total revenues from the surcharge plus the property tax (and other miscellaneous sources) are calculated to cover all its costs. However, because many firms' contributions to the district through their property taxes is thought to be higher than the cost of treating their wastes, industry as a whole is subsidizing residential users of the system.

The subsidy is increased by the industrial surcharge program since it enlarges industry's contribution to the district's revenues. This has generated some protests about the equity of the system, but there appears to be little concern about the accuracy of the unit prices calculated by the district for the surcharge program.

Impact on Industry

A wide range of industries is affected by the surcharge program in Chicago. Food processors are among the larger payers. The district has provided a computer printout, listing all payers and their wastewater flows and characteristics, real estate taxes, and surcharges for January to June 1978.

For all practical purposes, no firms within the district's jurisdiction are discharging industrial waste under an NPDES permit. A few firms have permits to discharge cooling water. This has been the pattern for years and there has been no noticeable change since initiation of the surcharge program.

Summary

The Municipal Sewer District of Greater Chicago has taken an aggressive role in the implementation and administration of its industrial waste surcharge program. Court ruling, district verification of self-monitoring, and determined enforcement in the intitial year of the program appear to have overwhelmed opposition to the surcharge program. With a few notable exceptions, it would appear that the surcharge has become a routine practice.

The surcharge program has been very successful at its stated objective of generating revenue, but it appears to have had little effect on industrial wastewater management. Any reduction in flow or BOD or SS levels stimulated by the program would be difficult to detect in such a large system with a growing number of users.

The imposition of the proposed user-charge scheme is expected to triple the number of firms subject to the surcharge regulations. At the same time, no one expects the revenues from the program to increase proportionately. District staff members are concerned that the revenues gained will not justify the increased administrative costs.

7 Dallas

Agency Overview

The Dallas Water Utilities Department is a division of the city government of Dallas, with independent financing through revenues. The utility serves the city of Dallas plus certain municipalities with which it has contracts. The surcharge system applies only to the firms within the city limits or to contract cities as a whole. Capital costs are financed through revenue bonds, grants, and contributions—usually in-kind—by developers for new services. Figure 7-1 shows a service area map. Figures 7-2 and 7-3 are organization charts of the Dallas Water Utilities Department.

The annual agency budget presents information on goals (table 7-1) and revenues (table 7-2), as well as expenses (table 7-3). According to the terms of its revenue bonds, the system is required to have an excess of operating revenues over operating costs equal to at least 1.5 times current-year debt-service costs. Any of the excess revenues not needed for debt service is to be placed into a special water and sewer reserve fund for financing the capital costs of future projects. The local shares of sewer construction costs and water-supply system capital costs are financed through both revenue bonds and accumulated reserves. Charge revenues are used for financing capital and interest payments on the revenue bonds. The fiscal year is October 1 through September 30.

Wastewater Management

Demand

As can be seen in table 7-4, water-supply services in Dallas have grown significantly in recent years, but there has not been a matching growth in wastewater flow, presumably because many of the new customers for water are using septic tanks rather than the sewer system for their wastewater. There has, however, been a steady growth since 1975 in industrial flow, which averages roughly 10 percent of total wastewater flows. Per capita wastewater flow is about 133 gallons per day.

Almost all water-supply and wastewater-discharge services are provided by the Dallas Water Utilities Department. Fewer than twenty firms supply

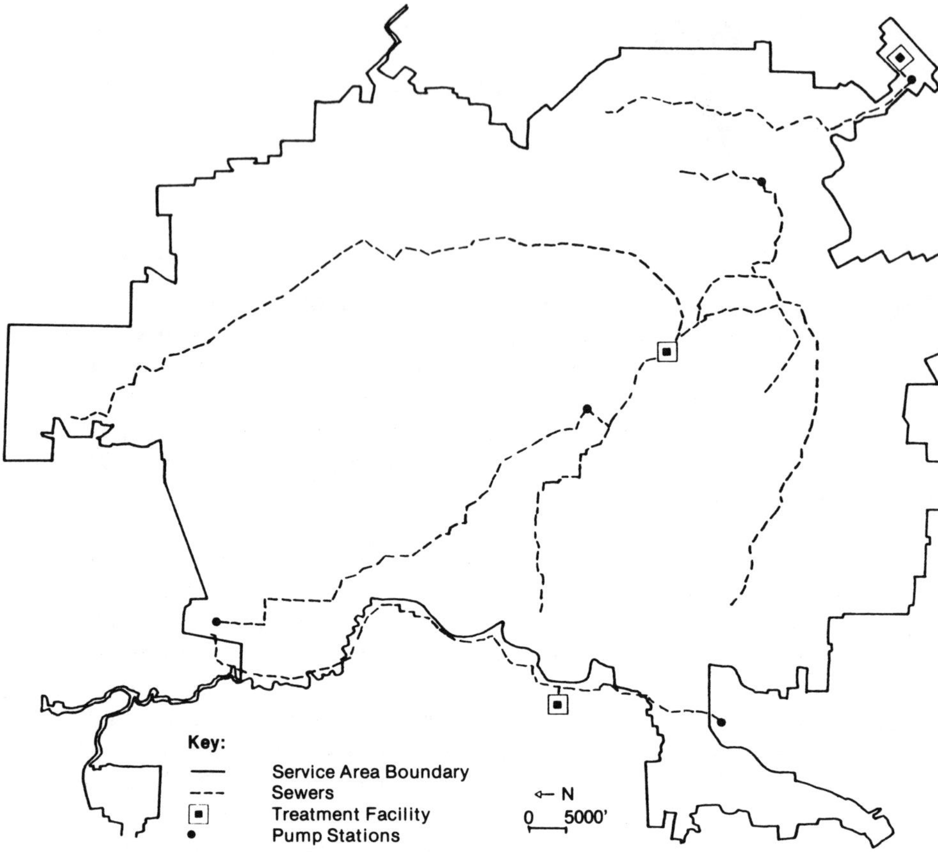

Figure 7-1. Dallas Sewer District

their own water either partially or completely, and even fewer (if any) firms have discharge permits. Self-supply of water is becoming less frequent, probably because of increasing costs; the level of self-supply is checked in site visits.

Supply Characteristics

The area has five treatment works:

1. The Southside Plant (4 mgd average flow; 53 percent removal of BOD and suspended solids; under expansion to 30 mgd with EPA Grant

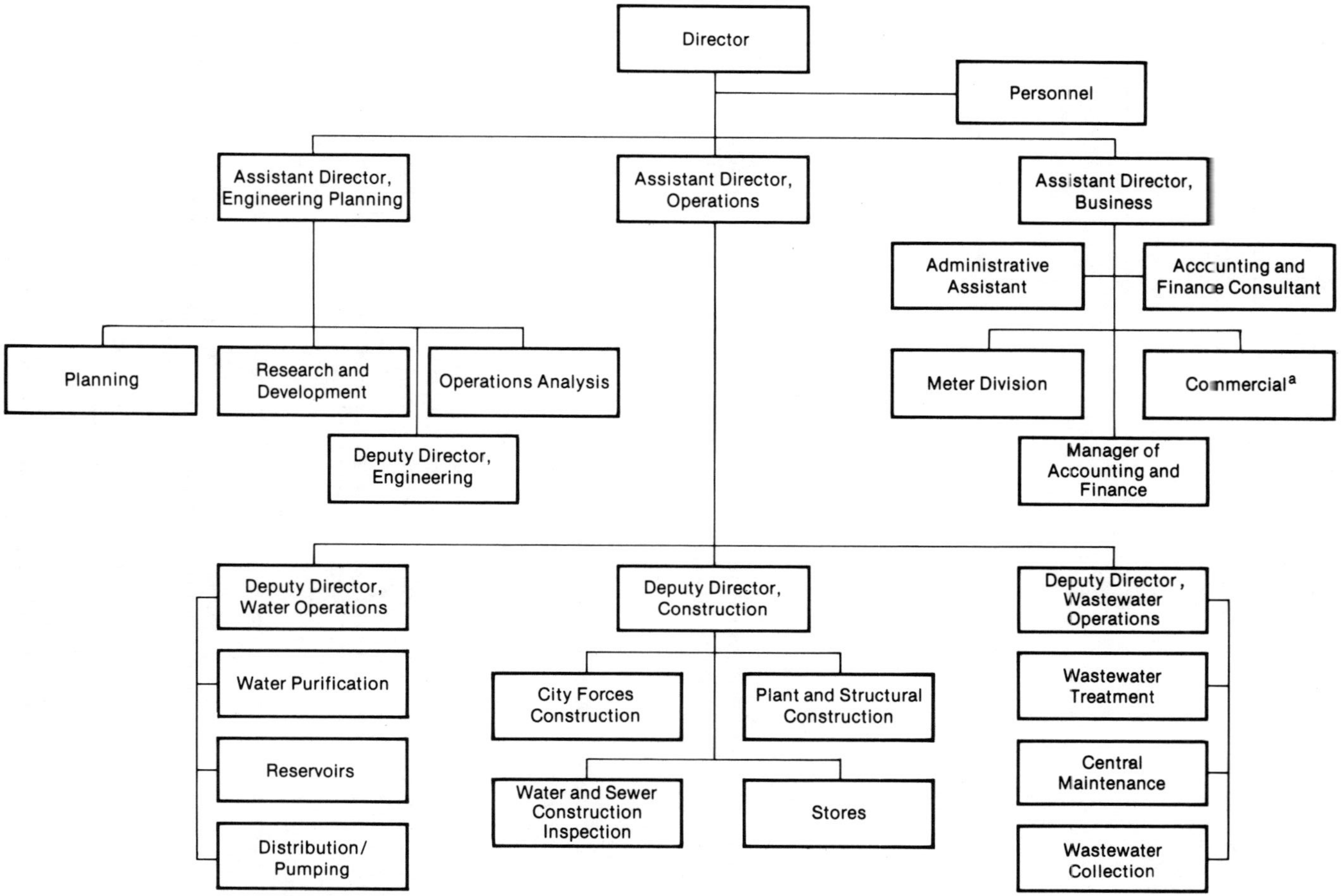

[a] See figure 7–3 for details of Commercial Division organization.

Figure 7–2. Dallas Water Utilities Department, Organization Chart, 1978–1979

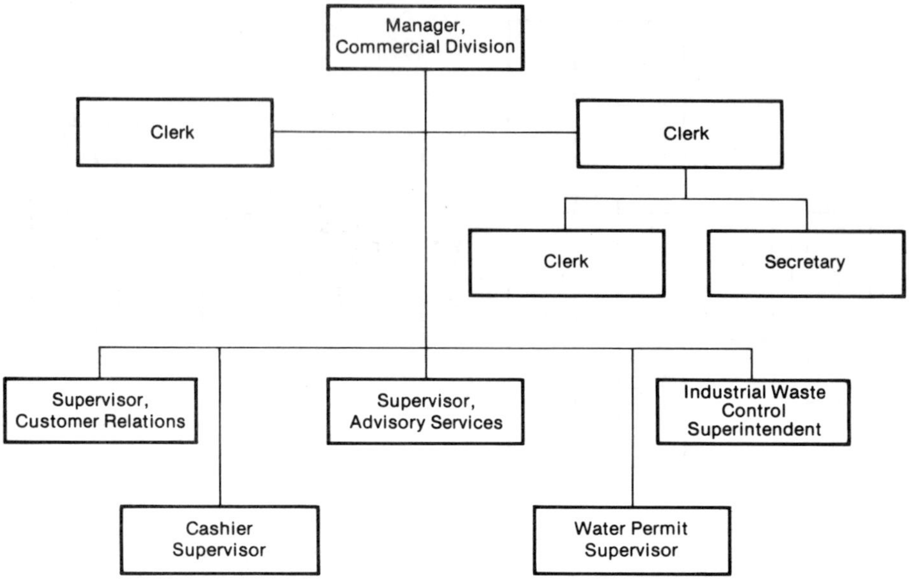

Figure 7-3. Dallas Water Utilities Department, Details on Commercial Division Organization

Table 7-1
Dallas Water Utilities Department, Goals

Program Description

The Water Utilities Department's program is designed to:

1. Provide an adequate supply of potable water that is free from health hazards, aesthetically acceptable, and adequate for household, commercial, and industrial use.
2. Collect and process sanitary wastes to assure that the City's wastewater does not adversely affect the environment.
3. Provide the services associated with potable water and with wastewater in a reliable, professional, prompt, courteous, safe and cost effective manner.

Program Objectives

1. Continue treatment processes which will insure high quality water for domestic and industrial use.
2. Continue the high level of removal of pollution in sanitary waste effluent.
3. Continue efforts to improve water and/or sanitary service to underserved areas.
4. Maintain an aggressive research program covering water and wastewater treatment and system operations.
5. Continue the task force approach to achieve increased productivity and cost reductions.
6. Continue efforts to achieve improved patron relationships with special emphasis on meetings with utility managers and officials of neighboring cities.
7. Devote special effort to improve departmental performance in areas of Affirmative Action, Safety, Personnel Attendance, and Career-Professional Development.

Table 7-2
Dallas Water Utilities Department, Revenues

Revenue Code	*Description*	*Actual 1976–1977*	*Estimated 1977–1978*	*Estimated 1978–1979*
9711	Treated Water Retail	30,025,460	30,702,311	30,582,000
9713	Treated Water Wholesale	4,544,749	6,044,966	7,936,000
9731	Raw Water Retail	212,167	171,457	200,000
9733	Raw Water Wholesale	112,815	107,158	120,000
9741	Water Service Charges	248,458	257,186	260,000
9742	Water Penalties	455,649	387,922	351,000
9743	Fire Protection	0	1,200,000	1,200,000
9751	Wastewater Retail Service	22,014,935	26,022,318	25,714,000
9753	Wastewater Wholesale Service	716,853	1,037,571	1,043,000
9755	Wastewater Surcharge	1,224,953	1,049,187	1,050,000
9772	Wastewater Penalties	299,308	234,924	210,000
	Total Utilities Revenues	59,855,347	67,215,000	68,666,000
9301	Interest on Bank Balance	5,252,419	5,780,000	5,300,000
9407	Federal Grants	16,042	0	0
9791	Sanitation Fee	346,000	363,000	381,000
9792	Miscellaneous	225,795	301,000	300,000
9952	Taps	445,629	727,000	750,000
9954	Water Pro Rata	284,307	391,000	400,000
9953	Laterals	244,959	427,000	450,000
9955	Sewer Pro Rata	458,843	425,000	450,000
	Total Revenues	67,129,341	75,629,000	76,697,000
	Net (Increase) Decrease for Stores	322,173	(138,000)	100,000
	Total Funds	67,451,514	75,491,000	76,797,000
	Less Total Expenditures and Transfers	64,780,924[a]	71,622,383[a]	76,090,008[a]
	Net Excess of Revenue over Expenditures			706,992

[a] Includes accruals and transfers at fund level.

C-48-1124-03; sludge piped in from Dallas/White Rock/Tertiary for treatment and disposal).

2. The Trinity River Authority Plant (not operated by the utilities department but receiving some flow under contract with it).
3. The Dallas Plant (34.5 mgd average flow; 85 percent BOD and SS

Table 7-3
Dallas Water Utilites Department, Statement of Financial Position, September 30, 1977 and 1976
(*thousands of dollars*)

	1977	*1976*
Capitalization		
Accumulated net revenues	$241,895	$223,386
Contributions in aid of construction	116,956	107,765
Long-term debt		
Revenue bonds payable, less current maturities	124,575	109,220
Water-supply and wastewater-service obligations, less current maturities	13,170	13,493
	496,596	453,864
Other noncurrent liabilities		
Construction advances	4,060	3,223
Customer deposits	3,512	3,262
	7,572	6,485
Current liabilities, payable from restricted and other assets		
Current maturities		
Revenue bonds payable	9,645	9,565
Water-supply and wastewater-service obligations	323	379
Accrued interest payable on long-term debt	1,296	720
Accrued vacation and sick leave	4,276	3,339
Accounts payable and other accrued liabilities	5,145	2,683
	20,685	16,686
	$524,853	$477,035

removal; trickling-filter secondary; on same site as White Rock and Tertiary).

4. The White Rock Plant (105.9 mgd average flow; 70 percent BOD and SS removal; high-rate trickling filters; on same site as Dallas and Tertiary).
5. The Tertiary Plant (140.4 mgd average flow received from Dallas and White Rock; activated sludge, which is operational, and multimedia filters; 95 percent BOD and SS removal from the influent received from the Dallas and White Rock Plants).

The vast majority of total flow is treated at the Dallas/White Rock/Tertiary complex, which discharges treated wastewater to the Trinity River and pipes sludge at about 1 percent solids to the Southside Plant. The complex appears to have some minor operational problems. It is aiming for discharge of roughly 10 ppm BOD and SS, but utility representatives claim that discharges last year averaged 8 ppm BOD and 15 ppm SS. In rough terms, the complex flow train involves grit removal, primary sedimentation, trick-

Table 7-4
Dallas Wastewater Demand

	1974	*1975*	*1976*	*1977*	*1978*
City area (sq. miles)	301.3	342.9	344.0	365.7	377.6
Number of water accounts	223,239	223,999	226,132	228,250	232,633
Connected sewer population (thousands)	945.5	963.5	981.5	996.5	1,013
Water supply (million gallons/year)	70,510	67,540	72,581	77,369	84,393
Wastewater flow (million gallons/year)	47,800	52,300	47,400	53,800	49,200
Industrial wastewater (million gallons/year)	5,800	5,100	5,400	5,500	5,700

ling-filter secondary treatment, activated-sludge secondary treatment, and multimedia filtration for part of the stream (although the multimedia filters are still being debugged). Effluent chlorination is performed. Sludge from the trickling filters is anaerobically digested, and waste-activated sludge is aerobically digested before transport. Metals do not appear to be a major problem, although sludge analyses were not obtained.

Figure 7-1 shows a sewer map. The sewer system is being expanded, with an estimated fifty miles of sewer main (or about 1.5 percent) added each year. Contributions from individuals—for laterals—and developers amounted to about $80 million per year in 1976 and 1977, with probably over half of that in-kind contributions such as collector sewers.

Nonpoint source management is not under the jurisdiction of the utilities department, although stormwater facilities may be constructed.

Industrial Surcharge Program

History

The industrial surcharge program, initiated in 1959 by a general sewer-use ordinance, was designed to cover the costs of treating industrial wastes with excessive concentrations of BOD or SS. The rates set in 1959 remained unchanged until August 1976, when the formula was revised and charges were increased, and rates were raised again in February 1977. (Rate levels for the different periods are described in a later section.) The 1976 revisions also strengthened the prohibited-waste requirements and initiated the commercial rates for small discharges of strong wastes, which could not be individually monitored.

Provisions

Dallas' current (1979) water and sewer rates are shown in table 7-5. For all customers there are fixed charges based on the size of their water meters. In addition, there are usage charges, with different rates for different classes of customers, and different summer and winter rates for residential customers. A few industrial customers with large amounts of water in their products elect to use wastewater meters and thus become eligible for special sewer rates; however, they pay the same water rates as other industrial customers if they don't obtain their water from a well. The optional general service rate is available for the largest water users, whose usage is relatively constant over the year, but only a few firms use this rate.

Two types of strength surcharges are applied in addition to these rates. The first is applied to commercial users, who are divided into three classes, each with a different surcharge rate:

1. Eating places, whose surcharge is twenty-eight cents per 1,000 gallons
2. Equipment service facilities, whose surcharge is fourteen cents per 1,000 gallons
3. Processors of food and kindred products, whose surcharge is twenty-four cents per 1,000 gallons

The approximately 2,700 commercial firms in these classes are not individually monitored but are charged according to the average wastewater strength of a sample of firms in the class.

The second type of strength surcharge is applied to industrial firms that have wastes exceeding 250 ppm in BOD and 250 ppm in SS and either cannot be included in a user class or request separate treatment from other class members.[1] Firms in this category are monitored individually, both to measure their BOD and SS levels and to determine if they have any prohibited wastes. The firms monitored for prohibited wastes are largely metal-finishing plants, which do not pay a surcharge because of their low BOD and SS levels. In 1978 185 firms were monitored by Dallas Utilities, but only 119 firms paid the industrial surcharge (see table 7-6).

The 185 individually monitored firms in 1978 contributed the following equivalents to the Dallas wastewater load:

Flow equivalent to 117,000 population.

BOD equivalent to 903,000 population.

SS equivalent to 338,000 population.

This was in addition to the demands placed on the system by the 1,013,000 population served by Dallas Utilities. Thus, monitored industries are

Table 7-5
Dallas Water Utilities Department, Net Monthly Rates
(dollars)

	Water	Sewer	Combined
Customer Charge			
5/8 inch meter	1.10	2.15	3.25
3/4 inch meter	1.20	2.15	3.35
1 inch meter	1.50	2.15	3.65
1 1/2 inch meter	2.00	2.15	4.15
2 inch meter	2.55	2.15	4.70
3 inch meter	17.65	2.15	19.80
4 inch meter	26.01	2.15	28.16
6 inch meter	51.35	2.15	53.50
8 inch meter	95.68	2.15	97.83
10 inch meter	149.58	2.15	151.73
Usage Charge (rate per 1000 gallons of water used)			
Residential			
Up to 20,000 gallons	.61	.38	.99
Above 20,000 gallons			
Winter	.61	None	.61
Summer	.79	None	.79
Summer Rate Billed in June, July, August & September			
General service[a]	.43	.41	.84
Optional general service			
1st million gallons or less	430.00 (minimum)	.41	.84
Above 1 million (per 1000 gallons)	.26	.41	.67
General service waste-water metered separately[ab]	.45 per 1000 gallons of wastewater discharged		

Note: Wastewater effluent containing chemicals or substances harmful to the sewage collection system or exceeding BOD demand and suspended solids constitute special situations and are billed separately after establishment of proper surcharge rate. Outside city limits, rates are multiplied by 1.5. Rates became effective 1 February 1977.

[a] Plus appropriate customer charge.

[b] If firm does not have its own well, it pays one of the general service rates.

responsible for about 10 percent of the wastewater flow but about 50 percent of the BOD and 25 percent of the SS.

As was indicated, the industrial surcharge rates were raised in August 1976 and again in March 1977. The rates in the different periods have been as follows:

1959–July 1976: \$.063 (BOD–325) + \$.063 (SS–325) per million gallons

August 1976–February 1977: \$.10 (BOD–250) + \$.08 (SS–250) per million gallons

March 1977–present: \$.20 (BOD–250) + \$.202 (SS–250) per million gallons

Table 7-6
Industrial Plants Individually Monitored or Surcharged by Dallas Water Utilities Department

Fiscal Year	*Plants Monitored*	*Plants Surcharged*[a]	*Average Number of Plants Surcharged Each Month*
1973	203	NA	NA
1974	221	NA	NA
1975	179	117	106
1976	187	NA	104
1977	182	NA	113
1978	185	119	111

[a] Number of plants surcharged at any time during the year.

(where BOD is five-day biochemical oxygen demand in milligrams per liter and SS is total suspended solids in milligrams per liter). Both of the rate increases were accompanied by letters notifying the plants included in the program. The commercial rates for classes of users are determined by applying the charge formula to the BOD and SS levels found in a sample of users in the class, developing a rate per million gallons, and applying this to water use.

The current charge rates were set by the city council as part of a full rate package. They are based on a study by Turner, Collie and Braden, Inc., and Arthur Anderson Co. The study projected system costs for 1977 and 1978 and allocated them among the components of service, flow, BOD, and SS, as shown in table 7-7. The costs, calculated on an average-cost basis, include a contribution to general city overhead developed from a previous study. After further analyses in the city government and some changes based on recommendations of the utilities staff, the new rates were implemented. Table 7-8 indicates the 1976 rates, the consultants' recommendations, and the 1977 rates actually selected. The differences were based on changes in accounting of costs, suggested by city budget analysts.

It should be noted that the surcharge rate applies to wastewater flow, which is difficult to measure. Dallas has chosen to use input water as the basis of the charge in most cases. In a number of cases where firms supply their own water, well flows are measured to provide an estimate. There are only a few sewer meters. The surcharge system is able, however, to consider consumptive water use by applying plant- or class-specific factors to metered water use. A large number of plants have their surcharges

Table 7-7
Surcharge Coverage

	Eligible for Surcharges		*Actually Surcharged*	
	Number	*Percent*	*Number*	*Percent*
Atlanta	200		66	
Flow		15		
Chicago	2,000		350	
Dallas	185		119	
Flow		10		
BOD		50		
TSS		25		
Salem	50		18	
Flow				15
BOD				38
TSS				19
South San Francisco			24	
Flow				13
COD				47
TSS				34

computed as: surcharge rate × water use × (1 − consumptive-use fraction), with the consumptive-use factor computed individually.

Monitoring

All monitoring is performed by the Dallas Utilities' Industrial Waste Control Section, although there are infrequent independent checks by outside firms paid by customers. A monitored industry is typically checked twice yearly, but larger firms may be monitored more often. During each monitoring visit several days of composite samples are collected by automatic sampling equipment. One sampling visit a year is paid for by the city, but all others are billed to the firm. Firms may (and do) request resampling when the values seem unusually high, but they must pay the costs of the resampling. Sampling costs are shown in figure 7-4. The laboratory tests are selected by the utilities department for each discharger, depending on the plant's processes. Approximately ten requests for resampling occur annually. Firms are required to provide a location for monitoring. They are notified about a week in advance to ensure that they provide a usable monitoring point and that they will be in operation. Dallas Utilities considers that a week's advance notice is not enough time for a plant to make a major shift in production plans; in addition, unannounced spot

Table 7-8
Dallas Water Utilities Department, Sewer Rates

	1976 Rates	*Consultants' Recommendations*[a]	*1977 Rates, Based on Staff Recommendations*
Monthly customer-service charge:			
Residential	1.65	.30	2.15
General service	1.65	.30	2.15
Optional general service	—	.30	2.15
Gallonage rates per 1,000 gallons (domestic strength):			
Residential	.29 (1st 8,000 gallons water used)	.46 (all water used)	.38 (applies only to 1st 20,000 gallons water used)
General service	.24 (all water used)	.50 (all water used)	.41 (all water used)
General service on optional rate	.264 (all water used)	.50 (all water used)	.41 (all water used)
Sewer separately determined	.29	.55	.45
Surcharge rates (per mg/l) over 250, per 1,000 gallons:			
Biological oxygen demand (BOD)	.00010	.000171	.000200
Total suspended solids (TSS)	.00008	.000197	.000202

[a] The consultants' stated gallonage rate for domestic-strength sewage is $.55 per 1,000 gallons of water (winter flow) for customers within Dallas. The staff applied the percentage of wastewater flow to metered water flow.

checks are sometimes made, particularly for dischargers of prohibited wastes.

Enforcement appears to be straightforward. Very few cases of nonpayment have occurred, and a 5 percent penalty is routinely assessed for late payment. A collection group follows up on delinquencies and visits the customers after three months; service is shut off about a month later, after notice has been given. However, shutoffs for continuing commercial and industrial accounts occur rarely, if ever. The industrial surcharges are billed independently of regular water and sewer charges. Sampling charges are also billed separately, but there appear to be no serious problems in collecting them. (The process differs somewhat, however.) Collection of the surcharges from delinquent firms requires less than one day a month of staff time.

Disagreements on the monitoring results for the surcharge have been

FY 1978–79 R–5

Date: ______________________ Billing Number: ______________

Name: ______________________

Address: ______________________

Time and Dates sampled: ______________________

Charges Authorized by Ordinance 15179, Section: 49–109.1 (b) Ordinance Violation

Item	Cost Each	Quantity	Total
COST OF SAMPLE COLLECTION:			
Composite Sample	$36.00		
Grab Sample	11.00		
Special Sampling per hour	11.00		
COST OF LABORATORY ANALYSIS:			
pH	2.00		
Suspended Solids	8.50		
BOD	14.50		
Grease	14.50		
Cyanide	12.50		
Arsenic	14.50		
Cadmium	8.00		
Chromium	8.00		
Copper	8.00		
Lead	8.00		
Mercury	14.00		
Nickel	8.00		
Silver	8.00		
Zinc	8.00		
COD	9.00		
TOC	8.00		
Pesticides (First)	35.00		
Pesticides (Additional)	15.00		
	Total Cost:		

If more than one pesticide can be analyzed from a single chromatogram, the charge will be $35.00 for the first and $15.00 for each additional pesticide.

____

Robert N. Dill Manager

Note: Costs are sampling fees charged to monitored firms.

Figure 7–4. Dallas Water Utilities Department, Sampling Fee Report

resolved fairly easily. If there are serious disagreements, the suggested approach would be to perform three analyses—by the city's lab, by the firm's lab, and by an independent lab—and use the city's or firm's result closest to that of the independent lab, with the loser paying the costs for all three samples. So far, there has been no need for this process.

The city's laboratory was set up as a demonstration research facility and claims to have good quality control, although quality-control procedures were not assessed during this field trip. Quality control on sampling is difficult to analyze. The use of long-term composites and fixed sampling points and procedures should improve sampling quality. Beyond that, good sampling requires judgment, and the inspectors in the section are experienced. The turnover is low: six of nine inspectors hired in 1973 are still in the section.

Impact on the Agency

Total service revenue in the wastewater accounts was estimated at $28.3 million for fiscal year 1978. Revenues from the industrial surcharge were $1.3 million (4.6 percent), and the commercial (or user class) surcharge provided $0.4 million (1.4 percent).

Billing, administration, and sampling costs for the surcharge program totaled about $425,000 (table 7-9). This represents approximately 25 percent of the surcharge revenues. However, $102,600 of the costs are attributable to the monitoring of prohibited wastes, and only $322,400 to the industrial and commercial surcharge programs. Thus only about 19 percent of the surcharge revenues are accounted for by administration of the surcharge program itself. Dallas Utilities' metering, accounting, and finance functions have direct costs of 6.5 percent of all revenues for accounts, not subject to the surcharge, but administration of these accounts is, of course, less difficult. The proportional overhead costs for the surcharge program, although higher than average, do not appear excessive.

The surcharge rates are designed to cover both the costs of administering the program and the excess-treatment costs of the industrial and commercial firms. Consultants were hired to determine the rates that would accomplish that, but as is shown in table 7-8, the rates finally chosen differed from those the consultants recommended. However, the 1977 industrial surcharge rates were higher than those recommended by the consultants. Presumably, therefore, the surcharge revenues do in fact cover the excess-treatment costs attributable to industrial and commercial firms.

The other impacts on the agency from the surcharge program appear to be minor. The program has accomplished its goal of improving cost allocation, but little cost reduction appears to have occurred. The pro-

Table 7-9
Costs of the Industrial Waste Division Programs in Dallas, Fiscal Year 1979 (*dollars*)

	Industrial [a]	*Commercial* [b]	*Other* [c]	*Total*
Billing/administration				
Salaries and benefits [d]	37,400	18,700	18,700	74,800
Vehicles and transport	—	—	—	—
Laboratory expense	—	—	—	—
Equipment and office	—	—	—	—
Personnel outside area [e]	2,000	4,000	1,000	7,000
Overhead (computer, etc.) [f]	11,200	6,400	5,600	23,200
Subtotal	50,600	29,100	25,300	105,000
Monitoring [g]				
Salaries and benefits	68,100	17,100	27,000	112,200
Vehicles and transport	9,700	2,400	3,900	16,000
Laboratory expenses [h]	66,800	16,600	26,500	110,000
Equipment and office	6,700	1,600	2,700	11,000
Overhead [f]	43,000	10,800	17,100	70,800
Subtotal	194,300	48,500	77,300	320,000
Program total	244,900	77,600	102,600	425,000

[a] Industrial surcharges on individually monitored firms.

[b] Commercial surcharges on firms billed by class of user.

[c] "Other" costs are primarily those for monitoring plants for prohibited wastes.

[d] Consultants estimate that 50 percent of salaries and benefits should be allocated to industrial, 25 percent to commercial, and 25 percent to other.

[e] Consultant's estimate for enforcement.

[f] Consultant's estimate from fiscal year 1979 budget; includes administration, general expense, personnel, and accounting and finance. Overhead rate is 28.4 percent, which may be high.

[g] Estimated at 60.7 percent industrial, 15.2 percent commercial, 18.5 percent prohibited waste, 5.7 percent other, based on total sampling hours.

[h] Net of $17,000 income from laboratory charges.

hibited-waste activities, on the other hand, seem to have had a more significant impact, by reducing discharges of some substances and tracing the cause of plant upsets back through the system to dischargers (which was done for one substance).

Impact on Industry

The major industries affected by the industrial program have been food processors, paper mills (from wastepaper), detergent and cleaning-product producers, and industrial laundries. The prohibited-waste program has affected platers, and the commercial program now covers truckers, car washes, and restaurants.

Pretreatment has increased, but apparently because of the prohibited-waste program and not because of the surcharge program. It is reportedly cheaper for the city to treat the wastes than for the plants to do so, unless products (such as fibers in the paperboard firm) are being lost. Pretreatment has been suggested as a method of reducing or eliminating charges, but little action has yet been taken to implement this suggestion. However, this could change when the Advanced Waste Treatment (AWT) plant becomes operational and costs—and therefore rates—rise. Sharply rising hazardous-waste disposal costs will provide opposing incentives for industries.

Housekeeping improvements have been implemented in many plants, according to district representatives. Reduction in water usage has not been affected to any great degree. This may change if the current drought continues and Dallas Utilities promotes conservation measures.

Summary

The system appears fairly close to ideal for our analysis. It covers dischargers in the right categories, has a fairly simple administration and enforcement procedure, and is not overly complicated by self-supply considerations. In essence, industries can respond to surcharges in a limited number of ways:

Challenge the calculations of Dallas Utilities.

Purchase extra monitoring by independent labs.

Conduct studies to indicate a greater water loss in production and thus to reduce the amount of wastewater flow on which firm is billed.

Improve housekeeping.

Shift production practices or products.

Implement pretreatment.

Move or shut down.

Pay the bills.

Use some combination of these.

Because of the apparent efficiency of the operation and the methods for rate-setting, billing, monitoring, and so on, it appears that such a system could be implemented widely. With respect to further data needs, it would be worthwhile to go along for a morning of sampling and to visit the lab. Aside from that, the necessary data are available in the interview notes and related documents.

Note

1. No class user has requested separate treatment, probably because of the monitoring charges discussed.

8 Salem, Oregon

Salem has a distinct set of problems related to industrial wastes and has developed a unique set of waste-strength charges that help it cope with these problems. The main problem relates to cannery wastes in the summer and fall. This is also the period when the assimilative capacity of the Willamette River is most limited. The city uses *demand charges,* which relate to each industrial plant's maximum use of the system's capacity during the preceding year, and *loading charges,* which relate more to operating costs. Both charges are based on flow, pounds of BOD, and pounds of SS discharged.

The Salem population lives for the most part in single-family residences on large lots. Commercial development is contained in the central downtown area and in small shopping areas. Key elements of the economy are canning and other food processing, paper manufacturing, and government. The population is projected to grow nearly 5 percent a year through 1980.

Agency Overview

> Salem provides water service in the city and supplies water, on contractual bases, to the Keizer and the East Salem water districts, the City of Turner, and the Orchard Heights Water Association. Water is also supplied to the City of Stayton under emergency conditions. Sewage collection, treatment and disposal is provided throughout the city as well as for the Keizer and East Salem sewer districts.[1]

Salem's Public Works Department has separate divisions for water, for sewers, and for sewage treatment (see the organizational chart in figure 8-1). These divisions have separate budgets but make use of one planning and engineering staff.

Salem does not have a municipal solid-waste-collection service; private operators are licensed to do this. The city neither encourages nor discourages the use of garbage grinders to dispose of food scraps through the sewer system.

The city of Salem maintains three funds to account for the fiscal activities of the water and sewerage divisions:

1. The Utilities Fund is used for all water and sewerage operating receipts and disbursements, including debt-service payments. Although sepa-

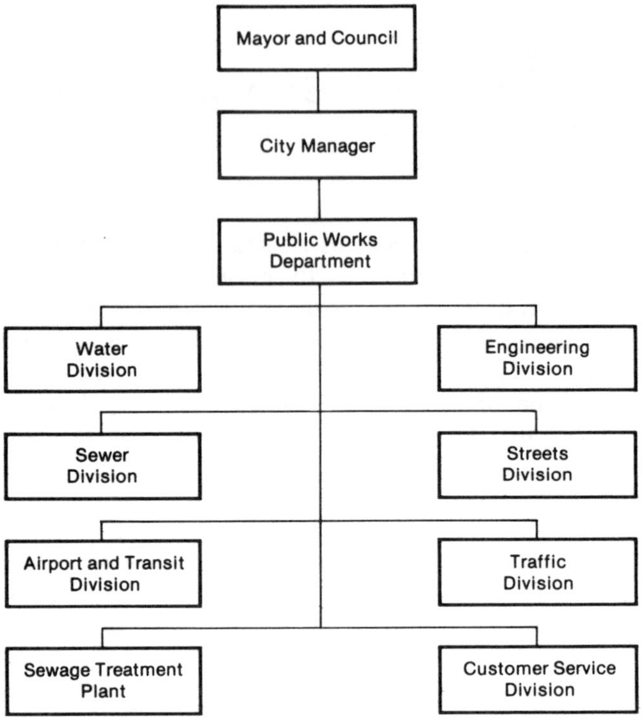

Figure 8-1. Salem Public Works Department, Organization Chart

rate accounting is maintained for income and expenditure by division, the monies on hand are equally available for water, sewer, and sewage-treatment operations. Any surplus accumulating in the Utilities Fund is available for capital expenditures.

2. The Bond Construction Fund is used to receive the proceeds from bond issues for designated water and sewerage construction projects and to account for the expenditures made for each project. Repayment of the bonds is accomplished through the Utilities Fund.
3. The Special Assessment Fund is used by all public works and utility divisions in the city to finance "improvement-district projects," that is, the extension of service to new customers. Subdividers and developers must either contribute in-tract water mains and sewers, or have the city put an assessment on the property to finance part or all of the in-tract facilities. Owners of assessed properties may pay the assessment in a single payment or in semiannual payments over a specified number of years. In the latter case, Bancroft Act improvement district bonds are sold to provide the long-term financing.

Both the water and sewer accounts are "enterprise funds," which means that the revenues accruing to the funds make them self-sufficient and independent of ad valorem tax revenue. The revenues are to cover all costs, including debt service. The revenues cover not only the administrative costs of the relevant portions of the Public Works Department but also the administrative costs of that section of the Finance Department that handles the funds pertaining to the sewer and water system. The system is also required to pay a specified percentage of the administrative costs of both the city attorney's office and the city manager's office, and these percentages are not reviewed very often to determine if they still accurately reflect the proportion of staff time devoted to business relating to the water and sewer system.

Water and sewer rates are periodically redesigned so that each of these services will break even over time. When a deficit occurs, a rate study is undertaken and new rates are instituted. As can be seen in table 8-1, a deficit occurred in 1974–1975, triggering the Brown and Caldwell rate study and new rates effective July 1976. Another deficit occurred in 1977–1978, so a new rate study has been ordered.

Wastewater Management

Demand

Customers are classified into several categories to isolate the various demands placed upon the utility systems. The classifications for the water and sewer systems are not the same. Plant-capacity requirements imposed by each class and annual usage volumes (as of 1975) are summarized in table 8-2. Because of the seasonal nature of the canning industry, the industrial sector is responsible for a far higher percentage of the plant-capacity requirements than of the annual load.

Supply Characteristics

The water system consists of slow-sand filtration facilities located at the North Santiam River, a transmission capacity of 70 mgd from the North Santiam source of supply to the city, a 100-million-gallon asphalt-lined reservoir located near the city of Turner, 20-million-gallon covered storage facilities in the city of Salem, and 323 miles of distribution lines, including thirteen lift stations. The city has rights to divert approximately 71 mgd and is negotiating with the corps of engineers for rights to an additional 65 mgd.

Some sewer lines dating back to 1881 are still in service in Salem. These lines carried wastes directly to the Willamette River. Waste treatment began

Table 8-1
Salem Utilities Fund, Comparative Statement of Annual Surplus (Deficits) and Working Capital Balance Changes
(dollars)

	Actual					*Unaudited*	*Budget*
	1972–1973	*1973–1974*	*1974–1975*	*1975–1976*	*1976–1977*	*1977–1978*	*1978–1979*
Revenues							
Water sales	1,577,907	1,529,877	1,579,965	1,892,067	2,443,951	2,705,413	3,431,905
Sewer service	1,696,233	1,780,037	1,839,035	2,684,872	3,952,089	4,133,267	4,522,835
Other	619,611	496,410	753,393	637,900	649,254	899,403	336,200
Total	3,893,751	3,806,324	4,172,393	5,214,839	7,045,294	7,738,083	8,290,940
Operating Expenditures							
Administration	754,067	958,918	1,023,879	1,160,447	1,289,951	1,114,021	1,864,890
Water operation	466,653	661,298	760,329	842,918	1,058,921	1,393,681	1,766, 775
Sewer operation	592,090	629,808	776,835	923,111	1,470,397	2,030,489	3,292,900
Total	1,812,810	2,250,024	2,561,043	2,926,476	3,819,269	4,538,191	6,924,565
Net Gain from Operations	2,080,941	1,556,300	1,611,350	2,288,363	3,226,025	3,199,892	1,366,375
Other Expenditures							
Debt service	1,053,197	1,071,185	1,189,933	1,453,512	1,463,207	1,357,762	1,841,980
Water construction[a]	351,998	325,801	422,104	353,610	145,397	1,088,194	661,788
Sewer construction[a]	252,152	156,186	256,339	204,980	220,795	862,253	661,788
Total	1,657,347	1,553,172	1,868,376	2,012,102	1,829,399	3,308,209	3,165,556
Annual Surplus (Deficit)	423,594	3,128	(257,026)	276,261	1,396,626	(108,317)	(1,799,181)
Beginning Working Capital	1,279,576	1,703,170	1,706,208	1,449,272	1,725,533	3,122,159	3,013,842
Total Annual Surplus (deficit) and Beginning Working Capital	1,703,170	1,706,298	1,449,272	1,725,533	3,122,159	3,013,842	1,214,661

Source: Salem Utilities Fund, August 15, 1978.

[a] Includes nonassessed and deferred costs from assessed projects.

Table 8-2
Water System and Sewerage System, Salem, Oregon

A. Water

Class	*Percentage of Plant-Capacity Requirements*	*Percentage of Annual Usage*
Single-family residences	42.6	37.2
Multiple-family residences	5.5	6.9
Commercial	12.2	15.8
Wholesale[a]	11.4	12.3
Industrial	23.1	22.7
Public (other than the city)	3.6	3.9
City use	1.6	1.2

B. Sewer

	Percentage of Plant-Capacity Requirements			*Percentage of Annual Load*		
Class	*Flow*[b]	*BOD*	*SS*	*Volume*	*BOD*	*SS*
Residential	24	7	15	31	31	41
Commercial	13	5	10	14	18	24
Institutional	6	2	4	7	12	14
Industrial	20	42	24	9	38	18
Infiltration/inflow	15	—	—	39	1	3
Future	22	44	47	—	—	—

Source: Brown and Caldwell rate study.

[a] Wholesale class is customers outside the city who purchase large quantities of water for resale.

[b] Flow that occurs during the summer canning season.

with the construction of a primary-treatment plant in 1953. Pressures in the early 1960s to clean up the Willamette River led to construction of the completely new Willow Lake Wastewater Treatment Plant in 1964. This trickling-filter system of secondary treatment removed about 80 percent of wastes, resulting in a treated effluent with average monthly levels of about 35 mg/l of BOD and 37 mg/l of suspended solids. Continued pressure to improve the river led to the addition, in 1976, of a parallel secondary-treatment system at Willow Lake, consisting of activated sludge with pure oxygen. The upgrading and expansion of the Willow Lake facility cost an estimated $18.5 million, of which 75 percent came from an EPA grant.

The expanded system has reduced average BOD discharges to about one fourth of their pre-1976 levels for the year as a whole and achieved even more dramatic summer reductions. Average BOD levels of 5 mg/l are common in winter months. Average levels of suspended solids in the treated effluent were cut in half, falling to 17 mg/l.

The sewerage system now consists of twenty-four lift stations, approximately 420 miles of collector sewers and interceptors, and two treatment plants. The Willow Lake Treatment Plant, which treats the bulk of the city's wastewaters, now has a capacity of 35 mgd flow, 142,500 pounds per day BOD and 76,000 pounds per day SS. Salem's second treatment plant, at Wallace Road, is a 0.04 mgd package plant constructed in 1969 to serve the north slope of West Salem.

The only major discharges to the Willamette River near Salem are the Boise Cascade Pulp and Paper Plant and the city waste-treatment plants. Because of its canneries and the high levels of BOD it has to treat, Salem is permitted to discharge to the river 11,000 pounds of BOD per day during the canning season. This represents 18 percent of the total permitted BOD for the river. In contrast, Salem has less than 10 percent of the valley's population. Improvements in the level of treatment are required, particularly if the city should decide to accommodate another large cannery. Conversations with two persons from Oregon's Department of Environment Quality (DEQ) and a 1975 DEQ study[2] indicate that the DEQ has concluded that land application of seasonal canner wastes is desirable. Canneries operate at a time of year when river flows and oxygen levels are critical and at a time when the land can absorb extra water.

High infiltration of water into sewer lines during excessively rainy weather means that the wastewater received by the Willow Lake Plant is sometimes extremely low in BOD. For example, the treated flows in December 1977 averaged 59.5 million gallons per day. With an inflow strength of 148 ppm and an effluent strength of 8 ppm, the treatment efficiency was only 87 percent. This presented no problems in the river, but the efficiency percentage may not conform to some EPA guidelines. Obviously, the solution is to reduce some of the infiltrations of ground water into the system, not to add more advanced waste treatment. The city has recently completed an infiltration/inflow study, which recommends a sewer-system evaluation survey to locate sources and quantify the cost effectiveness of rehabilitation alternatives. The maintenance budget is up from $400,000 in 1978 to more than $2 million per year for the 1978-1979 fiscal year. It is believed to be best to work away at this problem gradually and in conjunction with other projects of the Public Works Department.

Industrial Surcharge Program

Water Charges

Prior to 1975, Salem's water-consumption charges were based on sharply declining block rates, that is, lower rates for quantities above certain levels (see table 8-3). This was a holdover from an earlier era when the city may have been trying to attract industry. Currently there are only two quantity categories, and the difference between the rates for large and small quantities has been greatly reduced. There are also flat monthly fees, based on meter size, and separate capital charges for line extensions and capital improvements.

There is some question about the equity of these charges. According to Salem's revenue report for fiscal year 1978, customers classed as industrial bought 21 percent of the water and paid in 13 percent of the revenue. This does not necessarily mean that industrial customers were not paying their fair share of costs. No information was obtained on the cost of serving different classes of customers or on the seasonality of use. If industrial uses come mainly when storage facilities or system capacity are most limiting, then it is very likely that industry is not paying enough. Current rates, however, seem better than pre-1976 rates in billing industry equitably.

Table 8-3
Water Rates, Salem, Oregon, 1969–1979

Description	*Meter Size (Inches)*	*Monthly Quantity Allowance (Cubic Feet)*	*1969 to 12/31/75*	*1/1/76 to 12/31/77*	*1/1/78 to Date*
Minimum charge by meter size (dollars/month)	5/8, 3/4	400	2.00	2.50	2.80
	1	800	3.30	5.50	6.10
	1 1/2	1,600	5.90	11.00	12.20
	2	3,200	10.50	18.50	20.60
	3	6,000	17.50	34.00	38.00
	4	12,000	33.00	54.00	60.00
	6	24,000	64.00	115.00	125.00
	8	40,000	106.00	175.00	190.00
	10	60,000	158.00	240.00	270.00
Charges for additional water beyond the quantity allowance (dollars/ccf):					
1965 to 1975	Up to 2,000 cubic feet total		0.20	NA	NA
	2,000 to 100,000 cubic feet		0.13	NA	NA
	100,001 to 500,000 cubic feet		0.08	NA	NA
	Over 500,001 cubic feet		0.07	NA	NA
1975 to date	Up to 100,000 cubic feet total		NA	0.22	0.24
	Over 100,000 cubic feet		NA	0.16	0.19

Source: Brown and Caldwell rate study, table 1.

Some sixty firms and institutions are classified as industrial by the water division, whereas only fifty are so classified for purposes of sewer billing. The two divisions use separate definitions and billings. Very few industries have their own wells, but many homes have good well water and only use the city's sewer service.

Sewer Charges

Salem sharply increased its sewer rates 1 July 1976, when the expanded Willow Lake Treatment Plant went into operation. Rates for single-family residences were increased by 50 percent to a flat $4.50 per month, while rates for multiple-family residences rose 120 percent. Commercial rates were increased about 300 percent. The average increase for industries was about 200 percent, but the increase for some plants was over 560 percent. The main objective of the rate increases was to cover costs and to make industry pay its fair share (see table 8–4).

The Surcharge Program

The distinction between commercial and industrial users in Salem's ordinance is fuzzy, and it is sometimes advantageous for commercial firms to ask to be billed as industries. Industrial billing is based on wastewater characteristics of the user rather than on the nature of the user's business. A user may be excluded from the industrial category if the Public Works Department determines that he discharges primarily sewage or he introduces less than 50 pounds per day of BOD to the system. Users discharging marginally more than 50 pounds of BOD per day may be classified as either commercial or industrial, with the final decision lying in the hands of the director of public works. The department collects a twenty-four-hour composite sample to ascertain a user's BOD discharge level and periodically does spot checks to confirm that a firm is properly classified.

According to a city ordinance, plants discharging more than 200 pounds per day of BOD or SS must install an effluent measuring and sampling facility. Such plants, here labeled *monitored industries,* are monitored on a regular basis by Salem's Public Works Department. Salem has only eighteen monitored industries, although fifty firms are classified as industrial.

Industries or institutions discharging more than 10,000 gallons of wastewater per day must obtain a permit from the city and file a report that includes actual and predicted discharges. The city takes two composite samples per year from these firms to check the accuracy of their discharge reports and to determine whether their discharge is at a high enough level to

Table 8-4
Salem, Oregon, Sewer Rates Before and After Rate Increase of July 1976

		Rates per Billing Unit (Dollars)	
User Category	*Billing Unit*	*1965 to 7/1/76*	*7/1/76 to Date*
Residential			
Single-family	Dwelling unit	3.00	4.50
Multiple-family and mobile homes	Dwelling unit	1.50	3.30
Transients (hotels, etc.)	Rooms	0.75	Commercial
Commercial	Monthly minimum	2.25	4.50
	Loading charge beyond minimum, $/ccf	0.12	0.55
Institutional and industrial	Demand		
	Flow, mgd	714.90[a]	670.60[b]
	BOD, lb/day	0.105[a]	0.483[b]
	SS, lb/day	0.186[a]	0.114[b]
	Diameter-inch miles	None	21.90
	Number of accounts	None	2.36
	Loading		
	Volume, million gallons	None	79.94
	BOD, 1,000 lb	None	18.24
	SS, 1,000 lb	None	2.10

Source: Sewer Ordinance No. 1877 and amendments and Brown and Caldwell, "Salem Sewerage System Rates and Calculation of Industrial Wastes," tables 2 and 10-1.

[a] Charges were these amounts times the measured flow, BOD, and SS during the month of the preceding year that had the maximum load.

[b] Charges are these amounts times the individual parameter values for the month of the preceding year when that parameter was maximum. Since 7/1/76 there have been two reductions in these demand charges. The first reduction instituted a nonconcurrent usage factor to account for the fact that system peak loads are less than the sum of individual nonconcurrent peak loads. The second reduction came in 1978 when the U.S. Congress temporarily suspended the Industrial Cost Recovery Program.

require the installation of a sampling facility. Even if not required to do so by the city, these dischargers may install an effluent-monitoring facility and be classed as monitored industries. The rates, however, are the same for all industries, whether or not they are monitored.

The engineering firm that developed Salem's industrial charge program followed the cost-causative philosophy endorsed in 1973 by a joint committee of the American Public Works Association, the American Society of Civil Engineers, and the Water Pollution Control Federation.[3] The cost allocations used by the firm are indicated in table 8-5.

The new institutional and industrial charge formula that results from the rate study is described in table 8-4. The new rates have two components: a constant monthly demand charge and a loading or variables charge. The

Table 8-5
Cost Allocations Used to Determine Salem's Current Sewer Rates *(percentages)*

Annual Capital Cost Allocation	
Collection system:	
Maximum flow in a month	99
Diameter-inch miles	1
	100
Treatment plant:	
Maximum month	
Infiltration and inflow	32
Flow	16
BOD	46
SS	6
	100
Annual Operation and Maintenance Cost Allocation	
Collection system:	
Maximum flow in a month	35
Annual loadings	
Infiltration and inflow	4
Volume	7
Number of Accounts	52
Diameter-inch miles	2
	100
Treatment Plant:	
Maximum Month	
Infiltration and Inflow	12
Flow	6
BOD	25
SS	3
Annual Loadings	
Infiltration and inflow	11
Volume	23
BOD	12
SS	1
Number of Accounts	7
	100

Source: Brown and Caldwell, "*Water and Wastewater Revenue Requirements Study,*" pp. 46–47.

demand charge depends on the firm's monthly utilization of the system in the previous year, while the loading charge depends on the firm's system usage in the month in question.[4] A sample calculation is shown in table 8–6.

The loading charge is designed to recover industry's share of operating and maintenance costs, while the demand charges are based on capacity and capital costs.

Salem's earlier industrial charges (also shown in table 8–4) were based entirely on recorded effluents in the month of maximum BOD demand in

Table 8-6
An Example of the Calculation of an Industrial Wastewater Charge for the City of Salem, 1977

"Industry A," a hypothetical industrial user being billed for January, had the following discharges for the previous 12 months. The maximum-month discharges are circled. As shown, they occurred during different months.

Month	*Flow (mil gal)*	*BOD (lb)*	*SS (lb)*
January	1.02	4,250	3,400
February	0.86	3,500	2,900
March	1.51	6,300	5,000
April	3.20	13,300	10,700
May	4.57	19,100	15,200
June	5.21	21,600	17,400
July	7.50	31,300	(35,000)
August	10.22	(50,200)	28,200
September	(11.05)	38,000	31,600
October	2.60	10,800	8,500
November	1.36	5,700	4,700
December	1.00	4,100	3,600
January	1.50	6,300	5,100

This user's charge for January would be calculated as follows:

Month: January

			Parameter				
Line	*Description*	*Total*	*Flow*	*BOD*	*SS*	*Diameter-Inch Miles*	*Number of Accounts*
	Demand Charge						
1	Maximum-month discharge		11.05	50,200	35,000		
2	Maximum daily discharge (line 1 ÷ 30)		0.368	1,673	1,167	2.8	1
3	Nonconcurrent usage factor		0.729	0.858	0.538	1.000	1.000
4	Billing units (line 2 × line 3)		0.268	1.435	628	2.8	1
5	Unit rates, dollars		670.60	0.483	0.114	21.90	2.36
6	Demand charge, dollars (line 4 × line 5)	1,008.30	179.72	693.31	71.59	61.32	2.36
	Loading Charge						
7	Billing month discharge		1.50	6.3	5.1		
8	Unit rates, dollars		79.94	18.24	2.10		
9	Loading charge, dollars (line 7 × line 8)	245.53	119.91	114.91	10.71		
10	Total charge, dollars (line 6 + line 9)	1,253.83					

the preceding year. As such, they were more nearly capacity charges than loading charges. At that time, canneries were about the only monitored industries. They were monitored daily by the city while they were in operation. Thus, the system of industrial charges in effect from 1965 to 1 July 1976 did not provide a strong incentive to reduce flow, BOD, and SS, except in the month of maximum BOD demand. The addition of loading charges has altered this to some extent.

Monitoring

Each monitored industry is required to construct and maintain an approved measurement, composite-sampling, and sample-refrigerating facility for all waste flows entering the sanitary sewer. The flow-measuring device can be a Parshall flume, weir, venturi nozzle, magnetic flow meter, or any other device providing accurate and continuous flow indication. All samples must be continuously refrigerated at a temperature of 39°F, plus or minus five degrees.

The city is free to draw samples from these devices for analysis as frequently as it chooses. During the canning season, most plants are visited daily. Flows are recorded and samples are analyzed for BOD, SS, and pH.

The main cost of monitoring to the waste dischargers is the installation of the sampling equipment, estimated to cost about $6,000. The annual cost to the city of measuring the strength of industrial wastes and doing the extra billing is estimated to be roughly $50,000.

No major enforcement problems were noted. The city appears to have a good working relationship with firms. Regular surveillance of monitored industries is one of the main advantages of the surcharge system.

Equity

A comparison of industry's share of total wastes with its share of sewer revenues shows that industry provided 18 percent of all sewer revenues in 1977–1978 ($758,000 of $4,198,000). As of 1975, industry was contributing 9 percent of the flow, 39 percent of the BOD, and 18 percent of the suspended solids each year. During the month of peak industrial discharge, the corresponding percentages were 30, 42, and 24.

No information is available on the costs of servicing small versus large accounts. One thing is clear, however: Salem has a system in effect that allows it to bill offending industries for specific damages they might do to the system. Also, Salem has the desire to bill industry fairly, and the next round of rate adjustments will maintain this as a primary objective.

Impact on the Agency

The tasks of the Waste Treatment Plant Division are actually simplified by the industrial surcharge program that they must administer. It might seem that the extra administrative costs (estimated at roughly $50,000) are high compared to the total sewer revenue received from industry ($758,000 per year). However, the system of monitoring industrial plants makes possible systematic surveillance of industries and correction of problems before they become costly. Industries are supposed to have screens, settling ponds, and devices for neutralizing acid wastes. If these are not operating properly, the city can charge the offending firm for damages. But the problems rarely become that serious. Better surveillance means fewer repairs and better operation of the system as a whole.

The sewer ordinance specifies that industries must obtain permits to discharge wastes to the city system and that substances can be excluded if they are harmful to the system, the river, or the land on which the sludge is applied. The fact that the Public Works Department must administer a charge system that involves flow, BOD, and SS makes it clear to industry that the city is continually serious about other matters as well.

Impact on Industry

No evidence has been found to indicate that Salem's wastewater charges have caused industries to relocate. One firm went out of business for other reasons, and another enterprise took over that plant. Several new food-processing enterprises have located in Salem since industrial waste-strength charges began in 1965. Although there were only ten monitored industries in 1965, the number had grown to eighteen by 1977.

The wastewater-treatment plant supervisors are not aware of specific process changes or additional pretreatment that canneries have initiated on their own in response to waste-strength charges. Determining any such changes will be one of the main objectives of the second survey. The statistics in table 8-7 indicate that the seven canneries monitored continuously since 1965 have not significantly changed the BOD and SS levels in their wastewater.

Summary

The objectives of Salem's Public Works Department as it pertains to wastewater treatment are to provide good service at cost, to charge industry and other classes of customers fairly, to comply with the state's program for

Table 8-7
Wastewater Flow, BOD, and Suspended Solids for Seven Canneries Discharging Wastes to Salem Sewerage System since 1965

Year	*Flow (Million Gallons per Year)*	*BOD (Thousands of Pounds per Year)*	*BOD per Million Gallons Flow*	*SS (Thousands of Pounds per Year)*	*SS per Million Gallons Flow*
1965	341	2,711	7.95	1,260	3.69
1966	434	2,617	6.03	1,627	3.75
1967	506	2,839	5.61	932	1.84
1968	573	4,542	7.92	2,340	4.08
1969	538	3,275	6.09	1,871	3.48
1970	418	3,384	8.10	1,455	3.48
1971	404	3,467	8.58	967	2.39
1972	423	4,013	9.49	1,044	2.47
1973	562	4,839	8.61	1,797	3.20
1974	524	4,530	8.65	1,417	2.70
1975	473	3,494	7.39	1,314	3.00
1976	675	3,553	5.26	1,771	2.62
1977	587	4,441	7.56	2,483	4.23

improving the quality of the Willamette River by reducing the summer waste load, and to avoid subsidizing development. It appears that the city has a sewer ordinance and a set of charges that contribute effectively to the accomplishment of its goals. Charges are calculated to cover costs, and the industrial surcharges appear to be equitable to industry. There is, however, no evidence that the surcharges have reduced industry's demands on the sewer system.

Salem's industrial-waste charges rely about 60 percent on capacity or demand charges and 40 percent on loading or usage charges. Both depend on flow, pounds of BOD, and pounds of suspended solids. Capacity charges depend on the demands a firm made on the system in the months of maximum usage in the preceding year. Thus, firms have a strong incentive to reduce and even out their monthly demands on the system—particularly BOD. Loading charges, based on actual usage in the previous month, are incurred largely from flow and BOD. It is easily possible that monitored BOD discharges would account for half a firm's total sewer bill, combining capacity and loading charges.

Problems with the system appear to be mostly technical and pertain to infiltrations of water in the winter, unknown sources of suspended solids, and excessively acid wastes. High infiltrations and inflows of water into the sewer lines, especially in the winter rainy season, are long-run problems that Salem officials feel require a gradual approach rather than a crash program. The city and state need to agree on a strategy for replacing the worst lines and not let EPA impose costly advanced waste treatment as a way to cope with this winter problem.

The main problems on the Willamette River occur during the summer, when the river is low and canning industry discharges are high. However, the new treatment facilities have greatly reduced BOD and SS discharges into the river. There is no problem with BOD discharges in the winter months. In fact, the waste-treatment plant might have a higher efficiency ratio if it had more biodegradable wastes from December 1 through April 30.

Industrial dischargers are supposed to install and use devices for neutralizing excessively acid wastes. The city may start to charge for damages to the system when these devices are not used.

Notes

1. The rate study on which current charges are based is Brown and Caldwell, "Water and Wastewater Revenue Requirements Study, City of Salem, Oregon," Eugene, Oregon, July 1975.

2. "Water Quality in Oregon, A Status Report," Oregon DEQ, April 1975.

3. "The cost-causative approach incorporates the philosophy that a treatment works, by its very nature, must include certain components to achieve a stated level of treatment for normal conditions. The costs of these components are primarily related to their size. Accordingly, to recover costs in an equitable fashion, allocations to loading parameters are made on the basis of design sizing criteria. If a specific component is installed or oversized because of abnormal sewage loadings that require special handling, the cost of these facilities is automatically assessed to those users whose wastes cause the special design." Brown and Caldwell, "Water and Wastewater Revenue Requirements," p. 45).

4. The demand charge is reduced by a nonconcurrent factor, computed annually. This factor takes account of the fact that system peak loads are less than the sum of individual noncurrent peak loads.

9 South San Francisco

South San Francisco (SSFO) is known as "The Industrial City." Although SSFO has a great variety of industries, the city no longer has the many heavy industries (steel, slaughtering) that were located here earlier.

The city of SSFO includes large industrial and commercial areas in the eastern sectors adjacent to the Bayshore Freeway, bordering the San Francisco Bay or Colma Creek. These areas include the Utah Industrial Park and Cabot, Cabot and Forbes Industrial Park. Commercial activity in SSFO is centered around the downtown area (Grand Avenue) and along El Camino Real. The remainder of SSFO is primarily residential, as is San Bruno, which shares the SSFO wastewater-treatment facility.

Agency Overview

The city of South San Francisco has a history of innovative ideas and progressive programs for the treatment of wastewaters. SSFO was among the first cities to treat large amounts of industrial wastewater in a municipal system. In 1956 the city took the virtually unprecedented step of administering a special industrial charge for excessive suspended solids. This charge program has continued and evolved to the present.

This chapter focuses primarily on the system attributes of concern to the city of South San Francisco. These factors are not easily isolated, because SSFO historically has shared its treatment facilities with the city of San Bruno. SSFO independently administers the collection of wastewater from within its city limits and from two small abutting parcels, but it jointly owns, with the city of San Bruno, the subregional treatment plant that serves the two cities. This subregional facility shares a dechlorination plant and deep-water outfall with three local facilities, operated by the city of Burlingame, the city of Millbrae, and the San Francisco International Airport. Together, the four wastewater treatment plants, marked on the map in figure 9–1, are known as the North Bayside System Unit. This unit is commonly referred to as the *regional jurisdiction*. A schematic representation of the various elements in this regional jurisdiction is shown in figure 9–2.

The local SSFO service area includes the entire area within the city boundaries, plus small parcels in the town of Colma and in the Serrante-

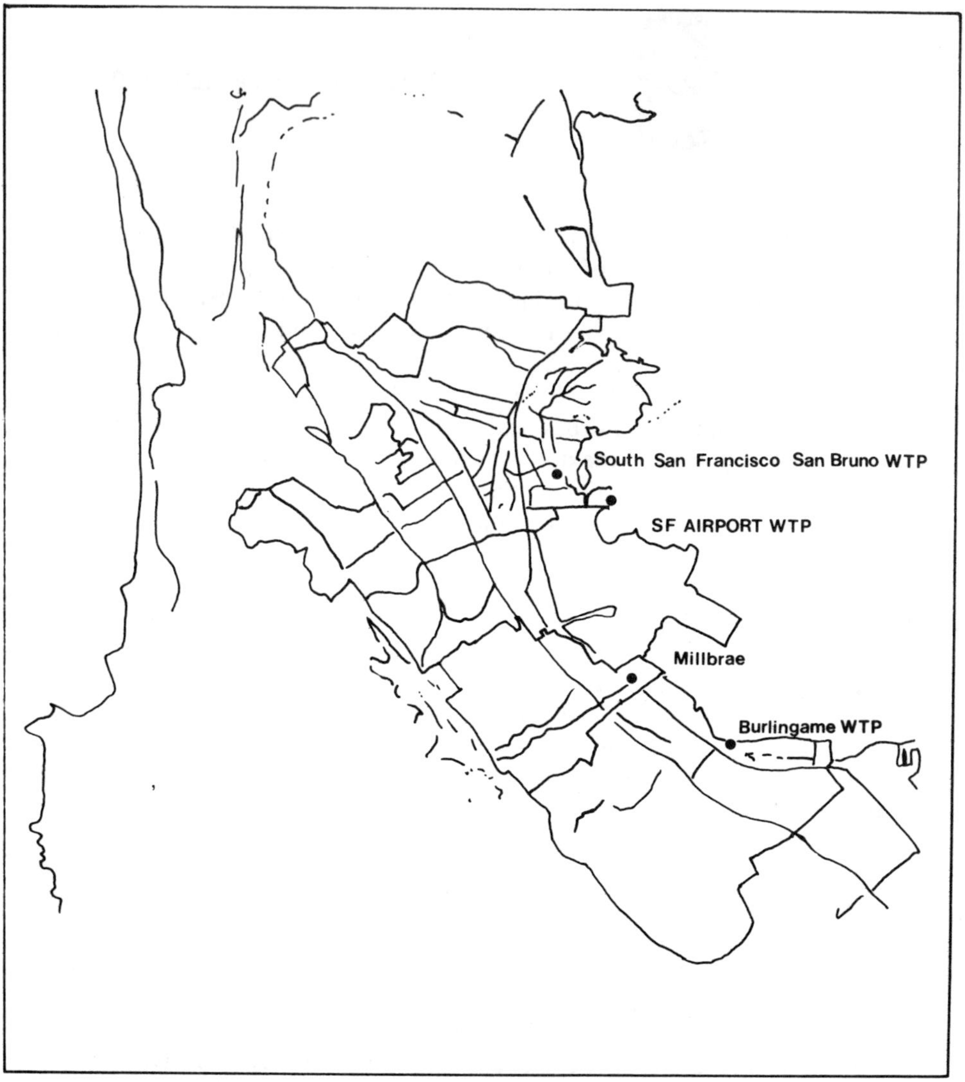

Figure 9-1. North Bayside System Unit Service Areas for Wastewater Treatment Plants

North San Mateo Community Sewer District. These small areas produce less than 3 percent of the effluent treated at the SSFO–San Bruno treatment facility.

The local SSFO service area has remained virtually unchanged since the

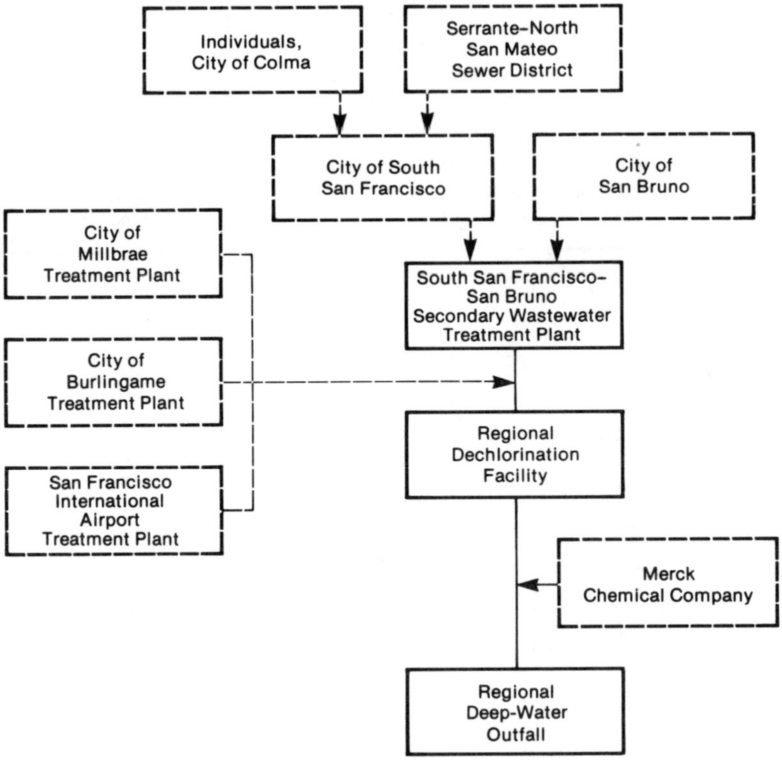

Figure 9-2. North Bayside System Unit

early 1950s, when the original treatment plant went into operation. The actual collection network has gradually expanded, and further service area expansions are anticipated.

The city of South San Francisco wastewater-treatment and collection facilities are both under the Department of Public Services. This agency is responsible for civil works, including roads, sewers, and wastewater treatment. The treatment plant and pump stations are run by the Wastewater Treatment Division of the department. The collection network, not including the pump stations, is run by the Public Works Division. The revenue and financial aspects are coordinated by a separate department, the Department of Finance. An organization chart of the individuals currently responsible is shown in figure 9-3.

The equivalent of five full-time employees are assigned by the Public Works Division to maintain the wastewater-collection network. In the current fiscal year, the cost of operation and maintenance for the system was roughly $135,000. Approximately twenty-five full-time employees are

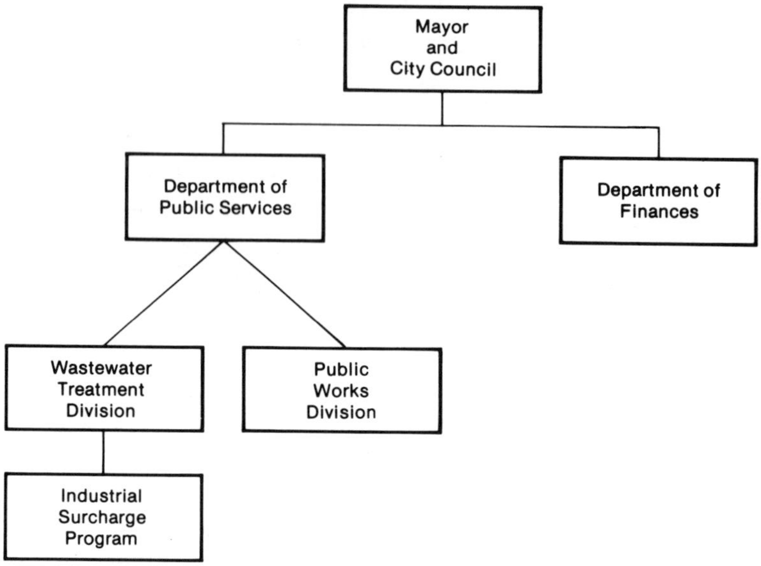

Figure 9–3. Organization of Wastewater Collection and Treatment in South San Francisco

involved in operations of the Wastewater Treatment Division. In the current fiscal year, the budget for this division was roughly $1.6 million.

The costs of collection and treatment, including capital, debt-service, and O&M costs, are recovered through user charges. Federal and state grants or loans are actively solicited for capital costs, however. Projected 1979–1980 costs and revenues are listed in table 9–1.

The costs of the subregional treatment facility are shared by San Bruno and the city of South San Francisco. South San Francisco owns the roughly ten-acre parcel on which the treatment plant is located. According to the joint powers agreement entered into at the time the treatment plant was being planned, the city of South San Francisco owns approximately 70 percent of the treatment works. The city also owns a share of a regional dechlorination facility, deep-water outfall, and the related transmission links. As the result of a variety of joint powers agreements, the city of South San Francisco operates and maintains not only the treatment plant but also the jointly owned effluent-transmission line, outfall line, and dechlorination facility. The city receives income from the participating public agencies and from Merck and Company, Inc., for the costs of facility operation and maintenance.

Table 9-1
Projected 1979-1980 Costs and Revenues for Sewerage in South San Francisco

Annual costs:[a]	
Operating and maintenance	$1,339,817
Debt	
Principal	
Interest	
Debt subtotal	231,359
ICR	34,542
Total	$1,603,718
Annual revenues:	
Residential, single-family	$481,911
Residential, multi-family	231,390
Institutional	37,611
Restaurants	45,322
Light-strength industrial	137,088
Moderate-strength industrial	145,022
Monitored industry	105,878
Monitored industry surcharge[b]	419,516
Total	$1,603,718

Source: Reta/Nolte and Associates, Consulting Engineers, "Revenue Program For Wastewater Treatment Facilities Serving the City of South San Francisco and the City of San Bruno," Final Report. Burlingame, Calif., 1978.

[a] Connection fees of approximately $130,000 are accounted separately into a capital-reserve fund.

[b] Based on COD and SS contribution.

Wastewater Management

Demand

According to a study prepared by Reta/Nolte Consulting Engineers, the SSFO-San Bruno Treatment Plant receives a wastewater flow of approximately 9.25 mgd. Table 9-2 indicates the average influent and effluent characteristics at the treatment plant during each month in 1975. Of the total waste load, roughly 59.2 percent of the flow, an estimated 74 percent of the total COD, and about 69 percent of the total SS loadings are generated in South San Francisco.

The concern of this chapter is with the South San Francisco wastewater contribution. South San Francisco has a population of roughly 45,000. About two-thirds of this population live in single-family dwelling units.

Table 9-2
Summary of South San Francisco–San Bruno Treatment Plant Performance for 1975

Month	*Average Flow (mgd)*	*BOD (mg/l)*		*Suspended Solids (mg/l)*	
		Influent	*Effluent*	*Influent*	*Effluent*
January	7.3	318	80	393	20
February	7.7	328	109	510	14
March	8.1	273	37	476	15
April	8.6	304	31	619	16
May	8.1	276	25	616	14
June	8.2	380	40	923	27
July	8.8	309	50	574	35
August	8.6	302	72	563	55
September	9.3	263	66	506	48
October	10.2	277	108	451	63
November[a]	10.4	234	86	225	69
December	8.8	295	78	194	40
Average	8.7	297	65	504	35

[a] The influent sampling point was changed to eliminate the influence of the centrate and supernatant.

There are approximately 1,400 business accounts in the service area, but only 25 are monitored and charged under an industrial surcharge program. The largest single source of wastewater is the residential sector, which produces an estimated 62 percent of the total flow. Only 12.5 percent of the total flow is produced by monitored industries (restaurants, institutions, and small business are responsible for the balance), but they are responsible for about half of the COD and over one-third of the SS. The estimated annual contributions from each sector, as reported in the Reta/Nolte study, are shown in table 9-3.

SSFO has experienced rapid growth over the past thirty years. In 1950, the city's population was approximately 20,000. By 1960, the population had doubled to nearly 40,000. Since then however, the growth has slowed considerably, and current population is estimated to be only 45,000. During the same period, the number of "wet" industries has declined. A discussion of this category is reserved for the section on the industrial surcharge program.

South San Francisco has a moderate amount of infiltration/inflow, which contributes to the amount of wastewater to be treated. During dry weather, total flows at the wastewater-treatment plant average approxi-

Table 9-3

Summary of Users and Wastewater Characteristics in South San Francisco

		Wastewater Characteristics			Design Capacity		Total Annual Capacity		
Users, User Groups	*Number of Users*	*Flow (mgd)*	*COD (mg/l)*	*SS (mg/l)*	*COD (lbs/day)*	*SS (lbs/day)*	*Volume (mg)*	*COD (lbs)*	*SS (lbs)*
Restaurants	70	0.083	1622	890	1123	616	30.2	0.410	0.225
Institutional	128	0.180	393	174	589	260	65.6	0.215	0.095
Small business									
Moderate-strength	230	0.324	1173	418	3172	1131	118.2	1.158	0.413
Light-strength	919	0.648	403	178	2178	962	236.5	0.795	0.351
Monitored Industry		0.683	2890	731	16473	4167	249.3	6.013	1.521
Residential									
Single-family	10070	2.277	403	178	7658	3384	831	2.795	1.235
Multi-family	4848	1.093	403	178	3677	1625	399	1.342	0.593
Subtotal		5.288	790	275	34871	12145	1929.8	12.728	4.433
Future capacity		0.2	270	90	450	150			
Totals		5.488	790	275	35321	12295	1929.8	12.728	4.433

Source: Reta/Nolte Consulting Engineers, "Revenue Program for Wastewater Treatment Facilities Serving the City of South San Francisco and San Bruno," 1978, p. 99.

mately 9 mgd, but in wet weather flows at the rate of 32 to 33 mgd have been observed. The SSFO portion of this excess flow is approximately in proportion to its total wastewater production. The city believes that illegal connections are a relatively minor problem. It should be noted, however, that at one time roof drains were legal, and some connections of this type may remain in the older parts of the city.

Supply Characteristics

The cities of South San Francisco and San Bruno have a long history of supplying wastewater collection and treatment. In 1945, the cities constructed a force main and pumping plant to convey septic discharges to San Francisco Bay and thus avoid discharging into local waterways. In 1951 and 1952 the cities jointly constructed their first wastewater-treatment plant, which provided primary treatment of wastes. Even today, industrial wastes are collected largely by a separate subsystem, because the original primary plant was designed to provide two independent treatment systems to allow for the separate processing of industrial wastes. This was done because there was some uncertainty as to the best practice for treatment of mixed municipal wastewater. At some early point, however, it was decided that combining the waste streams provided better treatment performance, and this practice has been followed since then.

By 1954 nearly all of the service area had been sewered, and the discharge of untreated wastewater had been eliminated. Increases in wastewater volumes and changes in the character of the wastewater to be treated led to a number of violations of discharge limitations imposed by the San Francisco Bay Regional Water Quality Board. To remedy this, the SSFO-San Bruno facility was upgraded to partial secondary treatment in 1962 and 1963. The original primary settling tanks were converted to aeration tanks. Vacuators for grease and flotable solid removal were added. Sludge incinerators, clarifiers, and a chlorine contract tank were also added. The design capacity of the secondary plant was 13 mgd, which was the average dry-weather flow from all the communities using the facility. Operating experience for the year 1969–1970 seemed to indicate successful performance of the facility, as shown by table 9–4.

By 1971, however, the wastewater-treatment facilities were faced with more stringent requirements regarding the toxicity of their effluent and the associated bacteriological characteristics. To meet these new demands, the focus switched from local and subregional facilities to a more regional perspective. A 1971 study recommended an additional aeration tank and digester and, on the regional level, an effluent pump station, outfall lines, and a switch from a shallow to deep-water outfall. These additions, completed in 1974, are shared by the cities of Burlingame and Millbrae, the San

Table 9-4
South San Francisco–San Bruno Treatment Plant Effluent Characteristics, 1969–1970

Item	*Actual Average*	*Discharge Limit*	*Plant Design Criteria*
Residual BOD (ppm)	108	200	—
Residual SS (ppm)	88	130	—
BOD removal (%)	74	—	62
SS removal (%)	84	—	79

Source: Jenks and Adamson Consulting Engineers, Inc., "Wastewater Management Program, South San Francisco–San Bruno Subregional Area Proposed Phase I Project Report," Palo Alto, Calif., November 1971, p. 2–5.

Francisco International Airport, and the Merck Chemical Company. The costs of the shared facilities are divided in proportion to peak flow contributions. Treatment-plant staff claim that the 1974 additions, especially aeration tank 7, have added greatly to the efficiency of plant operations.

The latest additions to the treatment works are currently being installed. In January 1977, a Jenks and Harrison study made two sets of recommendations for the plant. The first was for improvements in the SSFO-San Bruno treatment plant to bring it into compliance with full secondary wastewater-treatment-plant requirements. The upgrading, to be completed in 1979, includes 18 improvements, such as grit-changer improvements, return-activated sludge pumping and piping-system modifications, foam collectors on aeration tanks, and additional sampling and metering facilities. The second recommendation of the 1977 study was for regional facilities to dechlorinate effluent to meet a zero residual chlorine requirement. These facilities, completed in December 1978, use a sulfur dioxide-based process to dechlorinate the treated wastewater effluent before discharge. The facility is run by remote linkage to the SSFO-San Bruno Treatment Plant.

Adjacent to the SSFO-San Bruno Treatment Plant is a sludge-handling facility operated by Tillo Products, which has been producing compost since 1955. Originally, dried sludge was sold directly, but now Tillo markets the composted product. A new composting plant is currently being completed, which will mix sludge from the treatment plant, coffee grounds from the SSFO Folger's Coffee Plant, and rice hulls from other sources and turn them into compost. The new facility will have a capacity of 5,000 tons per month. There has never been a problem with compost disposal. The city was forced to build the new composting facility because the old site and operating procedures had contaminated areas of San Francisco Bay adjacent to the site.

Industrial Surcharge Program

History

Sewer charges have been levied by SSFO since 1951. In 1956, under City Ordinance No. 316, the first industrial surcharge was levied on concerns carrying out "manufacturing, producing, or processing operatings of any kind or nature." The charge was based upon flow and the concentration of SS. There was no clear provision as to how SS concentrations were to be determined. Quantity was based on metered water consumption, unless the user could offer proof that not all water consumed was returned to the sewers. In 1964, following a major expansion of treatment facilities, an amended sewer charge ordinance (No. 502) added a surcharge for grease in excess of 60 ppm. While the original ordinance had provided charges for suspended solids in excess of 500 ppm, the new ordinance decreed charges for excesses above 275 ppm. Again, no specific provisions were made for measuring SS and grease concentrations. Instead of the original flat rate for volume of wastewater flow, a declining block rate was introduced in 1964, but in March 1969, Ordinance No. 590 returned the charge structure to a flat rate for volume of wastewater flow.

In February 1972, a substantially different charge structure was introduced. The new surcharges for BOD and SS were based on the total amount of these discharges rather than on their concentrations. Surcharges were imposed only when the concentration was above 250 ppm, but the charges were based on the total number of pounds discharged. Again no provisions for monitoring of these discharges were incorporated.

In March 1974, a much more detailed program was implemented under Ordinance No. 662, establishing flow surcharges on suspended solids, by weight, when the concentration exceeded 250 mg/l; BOD, by weight, when the concentration exceeded 250 mg/l; and COD, by weight, when COD exceeded the BOD level in the wastewater discharged. The reason for a switch to a chemical oxygen demand (COD) test rather than a biological oxygen demand (BOD) test was twofold. First, the wastes produced by many SSFO industries are toxic. Therefore, the test may be inaccurate because the organisms depleting the dissolved oxygen are adversely affected by toxic components. Second, the sludge production, and therefore the costs of treatment plant operation, were not accurately predicted from BOD tests.

The March 1974 ordinance specified, for the first time, that the city take at least four test samples in measuring wastewater concentrations. It also specified that the procedures outlined in *Standard Methods* be used in the monitoring. The costs of laboratory analyses were to be passed on to the user.

In April 1975, under Ordinance No. 693-75, the charge on BOD was eliminated, but charges based on weight were instituted for COD in concentrations exceeding 500 mg/l. In addition, provisions for formal protests of charges were added to the ordinance. In 1976, the appeal procedures were slightly modified. Ordinance No. 735-77 eliminated the minimum concentration levels above which charges were levied. Ordinance No. 764-78 modified the rates but not the structure of the charges. The various charge structures and rates used by SSFO at various times are outlined in table 9–5.

Provisions

Since 1974, the enterprises subject to the industrial sewer rates have included manufacturers, processors, producers, laundries, photofinishers, painting services, packagers, and electric services. Enterprises are to be classed as industrial if their wastewater concentrations are above a certain level, but enterprises can be so classified only if the department elects to measure their concentrations. Thus there can be an element of arbitrariness in the classification of enterprises.

The sewer charges proposed for 1978–1979 are outlined in table 9–6. These provide for flat annual sewer fees for residential users and either a minimum charge or a volume charge for institutional and commercial users, including restaurants. The volume charge for industries is the same as that for institutional and commercial users, but industries alone are subject to surcharges for SS and COD. Unlike most of the other cities studied, SSFO's surcharges are based on the total weight of SS and COD discharges rather than on their concentrations.

In addition to the industrial surcharge, SSFO sets limits on the wastes that can be discharged into the municipal system. A weak source-control program was instituted in 1955 and replaced with a stronger one in 1974. The 1974 ordinance (No. 661) regulates BOD, COD, flow, grease, pH, sulfide, suspended solids, flammables, sand, cement toxicity, pesticides, and organics and heavy metals. Certain dischargers are required to obtain special permits, and their discharges are monitored.

Impact on the Agency

According to city officials, the charge program has enabled the agency to recover both the cost of administering the program and all the extra costs of treating industrial wastes (both capital and O&M costs). There apparently have been few complaints concerning the surcharge program. There are about twenty appeals a year from all classes of users. The city has no problems with collection because they can put a lien on property in order to recover unpaid charges. The city officials feel that they have developed an

Table 9-5
History of South San Francisco's Industrial Surcharge Rates

			Charge Per Unit				Minimum Concentration Above Which Charge Applies			
Date	*Ordinance Number*	*Flow ($/ccf)*	*COD ($/1,000 lbs)*	*BOD ($/1,000 lbs)*	*SS ($/1,000 lbs)*	*Grease ($/1,000 lbs)*	*COD (mg/l)*	*BOD (mg/l)*	*SS (mg/l)*	*Grease (mg/l)*
May 1956	396	.065			9.00				500	
April 1964	502	.11			6.00	7.00			275	60
December 1968	584	.11			6.00	7.00			275	60
March 1969	590	.22			8.00	9.00			275	60
February 1972	628	.22		20.00	15.00			250	250	
March 1974	662	.22	10.00	20.00	15.00		[a]	250	250	
April 1975	693-75	.34	23.00		50.00		500		250	
May 1977	785-77	.28	31.00		73.00					
May 1978	764-78	.31	34.00		74.00					
March 1979[b]	—	.31	34.00		74.00					

[a] In excess of BOD.

[b] Proposed, not approved as of February 1979.

Table 9-6
City of South San Francisco, Sewer Rates, 1978-1979

Residential class	
Single-family	$43.00/year
Multi-family, per unit	43.00/year
Trailors, etc.	35.00/year
Institutional class	
Minimum charge, or	$37.00/year
Volume charge	0.37/ccf
Commercial class	
Offices	
Minimum charge, or	$37.00/year
Volume charge	0.37/ccf
Restaurants	
Minimum charge, or	$37.00/year
Volume charge	0.78/ccf
Industrial class	
Minimum charge, or	$37.00/year
Volume charge, and	0.31/ccf
SS charge, and	0.074/lb
COD charge	0.034/lb

Source: Ordinance No. 764-78.

equitable program and that the costs of monitoring and enforcement are small and are appropriately borne by the users.

Impact on Industry

The impacts on industry are difficult to delineate in a brief study. Both city and industry representatives agree that no industries have left because of sewer charges, although a number of industries have left SSFO or changed their operations considerably. These include Swifts, Spice Island, and Bethlehem Steel.

The consensus seemed to be that municipal treatment, at least before source control, was cheaper than self-treatment. In fact, no industries in SSFO do not take advantage of the municipal treatment for their wastes. The surcharge program has never been considered a means of stimulating changes in the quantities or types of wastewater generated; its goal has been simply to recover costs and to match charges with the burden a user imposes on the system.

Table 9-7 lists the plants monitored by SSFO in 1977-1978 and their estimated 1979-1980 wastewater contributions. Table 9-8 indicates the total wastewater discharged by monitored industries from 1969 to 1976. Since

Table 9-7

Estimated 1979–1980 Wastewater Contributions of Individual Plants Monitored for South San Francisco's Industrial Surcharge Program

	Wastewater Characteristics			*Design Capacity*		*Total Annual Capacity*		
Users, User Groups	*Flow (mgd)*	*COD (mg/l)*	*SS (mg/l)*	*COD (lbs/day)*	*SS (lbs/day)*	*Volume (mg)*	*COD (lbs)*	*SS (lbs)*
Armour	0.081	1,674	410	1,132	277	29.6	413,355	101,235
Berkey Photo	0.060	819	49	411	24	21.9	149,940	8,925
Berkeley Farms	0.059	3,620	1,394	1,790	690	21.7	653,310	251,685
Coast Linen	0.071	1,036	313	618	187	26.0	225,420	68,085
Dobbs	0.017	1,299	295	182	41	6.1	66,555	15,045
DuPont	0.007	4,189	295	249	17	2.6	91,035	6,375
DuBuque	0.045	3,263	377	1,211	140	16.3	442,170	51,000
Fuller-O'Brien	0.043	2,195	580	799	211	15.8	291,465	77,010
H.B. Fuller	0.001	8,710	4,114	124	59	0.5	45,390	21,420
Georgia-Pacific	0.032	6,240	2,390	1,622	621	11.5	592,110	226,695
General Printing	0.007	3,009	244	181	15	2.6	66,045	5,355
Lucca Packing	0.043	1,220	294	442	107	15.8	161,415	39,015

Maryatt Ind.	0.013	3,588	1,089	393	119	4.8	143,310	43,605
Orowheat Baking	0.010	8,403	4,130	711	349	3.8	259,590	127,500
Oppenheimer Casing	0.005	4,370	1,254	175	50	1.8	64,005	18,360
Poetsch & Peterson	0.037	3,995	832	1,230	256	13.5	448,800	93,585
Golden Gate Produce	0.010	492	205	42	17	3.8	15,300	6,375
Reichhold Chemicals	0.028	10,830	2,273	2,510	527	10.2	916,215	192,270
Spun Steak	0.002	791	326	13	6	0.8	4,590	2,040
Quong Hop	0.010	8,823	2,459	730	203	3.8	266,475	74,205
Columbus Salami	0.018	3,236	630	481	94	6.6	175,695	34,170
See's Candies	0.013	5,278	725	571	78	4.8	208,590	28,560
Sonoma Mission	0.018	279	54	42	8	6.6	15,300	3,060
Folger Coffee	0.050	1,039	187	811	70	10.4	296,565	25,500

Source: Reta/Nolte Consulting Engineers, *Revenue Program for Wastewater Treatment Facilities Serving the Cities of South San Francisco and San Bruno*, 1978.

Table 9-8
Wastewater Discharged by Industries Monitored in South San Francisco, 1969-1976

		Total Water Used by Sampled Industry (mgd)		*Total Wastes Generated by Sampled Industry (millions of lbs/year)*			
Year	*Number of Plants Sampled*	*Metered*	*Estimated Discharge*	*COD*	*BOD*	*SS*	*Grease*
1969	18	1.04	.85			2.71	1.51
1970	20	1.06	.87			2.60	2.06
1971	22	1.16	.95		2.84	2.97	
1972	22	1.14	.95		3.92	3.09	
1973	24	1.23	1.02	8.31	4.21	2.64	
1974	27	1.19	1.00	7.54	3.34	1.61	
1975	26	1.12	.90	7.28		1.88	
1976	23	1.03	.78	5.82		1.41	

Note: Statistics on waste discharge are based on grab sampling at a relatively low frequency (two to four times per year), supplemented by industry-supplied data.

1973, there appear to have been dramatic decreases in these discharges in terms of flow, suspended solids, and COD. However, the reductions in flow are not necessarily real reductions but may simply reflect "proofs" by industries that they are not discharging their total water consumption into municipal sewers. Also, California experienced a serious drought in 1976–1977, and industries in the South Bay were asked to reduce their water consumption by 25 percent. The reduced size of COD and SS discharges may be calculated simply from the apparent flow reduction. However, between 1973 and 1976 the COD concentrations from monitored industries appear to have fallen from .022 pounds to .020 pounds per million gallons of wastewater, and the SS concentrations appear to have fallen from .007 pounds to .005 pounds per million gallons (see table 9–6). It is difficult to tell how much of this reduction is due to the industrial surcharge program and how much to the source-control ordinance.

The source-control program, unlike the surcharge, has forced several concerns to do pretreatment. Certain provisions of the ordinance have also aroused the opposition of the Peninsula Manufacturers Association, an industry group. The SSFO's treatment plant people have found that the pretreatment or process changes triggered by the source-control program have reduced the waste loadings that the city must treat. The source-control program seems to have had far more impact than has the industrial surcharge. It is also the source-control guidelines rather than the industrial surcharge that has aroused strong opposition from local industries.

Summary

The city of SSFO has a long history of effluent charge programs, specifically industrial surcharges. These programs have been directed solely at the equitable recovery of costs. There is every indication that care has been taken to insure fairness and that the programs have succeeded with relatively little opposition.

Note

1. These estimates of current costs, provided by the Department of Public Services, are somewhat lower than the projected 1979–1980 costs in table 9–1.

Part III: Selected Industry Case Studies

This part reviews the industrial response to wastewater charges at 8 of the 101 case-study plants in the five cities. It includes examples from the major categories of facilities and shows the various types of responses described in chapter 3. Thus, it provides detail on the actions in particular situations, to supplement the summaries of responses provided earlier.

10 Tortilla Manufacturing

Introduction to the Plant

The firm is one of the largest of the approximately 400 firms that manufacture tortillas and taco shells to supply the rapidly growing Mexican-food industry in this country. This is the only location of the eight-year-old privately owned firm. The production of the firm has doubled over the last three years, with a concurrent doubling of production personnel. Of the 105 persons currently employed by the firm, 80 are directly involved in the production of approximately 12 million tortillas per month.

The Mexican-food industry in this country is rapidly growing. Although most firms producing tortillas are located in the southwestern portion of the country, predominantly in Texas and around Los Angeles, increasing demand for tortillas in the north, southeast, and northeast has broadened the market potential for existing firms. With an increased market, significant numbers of larger food producers have begun to manufacture tacos and tortillas. Perhaps the best-known popularly marketed items are tortilla chips, such as those manufactured by Frito-Lay. As larger firms enter the field, the actual number of plants will probably be reduced.

Although the potential market for tortilla and taco products is large, the firm focuses primarily on a single market sector, the private grocery purchasers. The product is sold in the dairy case, has a shelf life of 90 days, and is marketed through brokers on a guaranteed buy-back plan. If the grocer does not sell the product within 90 days, the firm pays both return shipping and retail cost. Losses due to no sale are almost nonexistent. The firm's products are originated and distributed through brokers to supermarkets throughout a twenty-state region. The firm is continually investigating potential for new marketing regions.

Production, Operations, and Physical Plant

The firm has three major products: corn tortillas, flour tortillas, and taco shells. All three products are sold in packages of a dozen. A three-shift, twenty-four-hour-a-day production schedule is required to produce the approximately 12 million tortillas and tacos per month.

In past years, competition for dairy-shelf space during the holiday season has reduced the orders for the firm's products. However, in the Christmas 1978 period, the reduction was not significant. The firm's management considers this to be an indication that the demand for its products has become more stable.

The firm's plant consists of two production lines: the corn line, which is used to produce corn tortillas and taco chips (figure 10–1), and the flour line, to produce flour tortillas (figure 10–2). Wastewater production from the flour line results primarily from a weekly washdown. Wastewater generated from the corn line has a significant production-water component as well as a daily washdown-water component.

The firm purchases its corn from a single farmer at a negotiated fixed price, thus reducing cost variability for this important input. Once received, the corn is stored until it enters the production process. The first step in the process is cooking in a water solution to which lime has been added to help loosen the corn husks. The corn is then steeped in a solution of water and sulfuric acid to help soften the kernels for milling. The corn is then washed and ground before being mixed with other ingredients to make the tortilla dough. Waste-process water from the cooking, steeping, and washing steps flows directly to the city's sewers with no effort to recycle.

The tortilla dough is then cut into the proper shape and baked. This is the end of the production process for corn tortillas, which are then packaged and stored for shipment. If taco shells are desired, an additional process step of frying is required prior to packaging and storage.

Housekeeping of the corn line requires daily washdowns. The plant management estimates that approximately one-tenth of the approximately 600 thousand gallons of city water used by the firm monthly is used in the housekeeping operation. Since the flour line is only washed down once a week, housekeeping on the corn line accounts for most clean-up water discharged to the sanitary-district sewer.

The flour line (figure 10–2) is significantly simpler. Flour, water, and other ingredients are mixed to form flour tortilla dough. The dough is then cut and baked, and tortillas are packaged and stored for shipment. Because of the inherent difficulties in mixing flour and water, the flour line is dry-cleaned on a daily basis and is scrubbed down with water only once a week.

Under current manufacturing practices, minimal solid wastes are generated by the production process. Corn husks are sold to a waste-food processor for use in animal feed. Similarly, the production process does not generate airborne emissions.

While the overall production processes have not changed, the firm has had a significant upgrading of capital equipment during its eight-year life. Because of rapid advances in technology, equipment purchased as recently as a year ago is in many ways obsolete. This rapid advancement in tech-

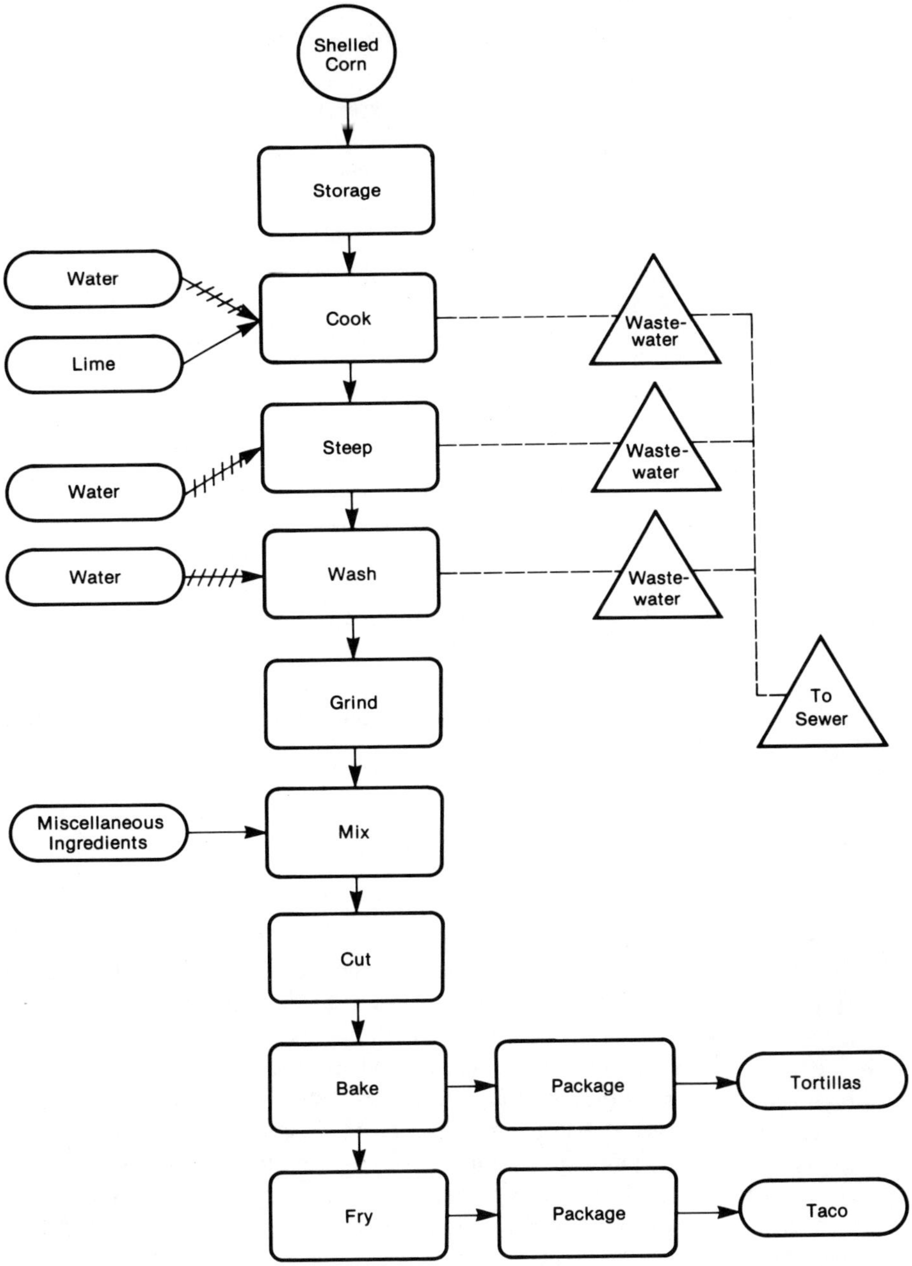

Figure 10-1. Corn Line

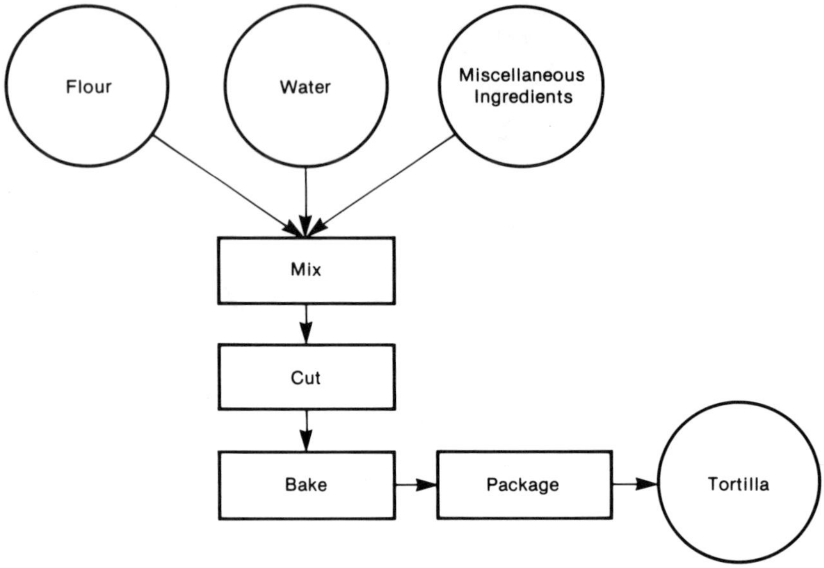

Figure 10–2. Flour Line

nology enables the firm to produce increasing volumes within a limited space.

Decision-Making and Response to Charges

Table 10–1 summarizes the firm's surcharge-payment and sampling history. These figures indicate that the surcharge payments have increased by 122 percent over the three-year period. When corrected for changes in the rates charged by the district, this still represents a large increase. The firm's water use has increased over the period. The firm has had only two samples analyzed under the program. However, the results indicate a wide variation, which cannot be attributed to any specific procedure in the plant.

The firm actively strives to minimize costs. Costs of all inputs—ingredients, energy, labor, packaging, and so forth—are monitored. Even though the costs of water and wastewater treatment are relatively small in comparison to these major inputs, they are consistently monitored as well. The firm is well aware that the surcharge exists and is continually reexamining the cost effectiveness of providing various degrees of pretreatment as opposed to paying the surcharge. The operations manager and the plant engineer have been primarily involved in this decision-making process.

Table 10–1
Surcharge and Sampling Data

Year	*Surcharge Payments*	*BOD (thousand lbs)*	*SS (thousand lbs)*	*Q (million gals)*
1978	$2870	14.1	49.4	6.9
1977	1779	16.7	35.7	6.0
1976	1296	23.6	20.9	5.23

Sampling Date	*BOD (mg/l)*	*SS (mg/l)*
5/78	199	870
1/77	541	484
1/76	541	484

In order to comply with the pH regulations, a new cooking facility was installed at the plant at a cost of approximately $30,000. The new cooking facility allowed the firm to use smaller volumes of lime. Pretreatment facilities, including a holding tank with pH adjustment, were also installed, to be used in the case of accidental discharge. Athough the manager feels they were forced to upgrade somewhat sooner than they would have, in retrospect the installation of the new cooking facility is considered to have been advantageous to the firm.

Approximately two years ago, the district discovered the existence of the firm and arranged to monitor effluent discharges. This monitoring indicated that the firm was in violation of both the limited-substances regulations, because of high pH, and the surcharge provision.

The firm's response to the surcharge has been to investigate the costs of installing pretreatment facilities, which would reduce or eliminate the surcharge fee. An engineering study concluded that, due to limited space for the required underground construction, installation of extensive pretreatment facilities would be too expensive, in the range of $25,000 to $40,000. The firm has considered the installation of screening devices, which would reduce the solids content of their discharge and significantly reduce the surcharge. However, space has again been the deciding factor against such projects. Even if sufficient space were available to physically locate screening devices, storage space would be needed for the solid-waste residuals. Storage of such organic materials in a food-processing plant presents significant sanitary problems. Organic solids generated by such pretreatment devices would presumably be marketable to waste-food processors.

The management of the firm feels that surcharges, which provide individual firms the option of paying a centralized district for providing additional treatment or providing their own treatment, whichever is economical to the individual firm, is a good system for providing environmental protection. This is based on the assumption that the charges are calculated on actual costs of operation to the sanitary district and are not designed to penalize or regulate industries. In general, the firm would like to concentrate on making tortillas and leave wastewater management to the sanitary district to the largest degree possible. However, paying for a "right to pollute" seems inappropriate to the firm.

Thus, the response of this plant to the surcharge has been to examine the alternative of pretreatment versus paying the charge, with no modification in production. To date, this analysis indicates it is to the firm's advantage to simply pay the surcharge. In addition to the direct costs of the surcharge, the administrative cost for sampling, filling out forms, and paying bills is barely significant. These costs include a fee of $250 per year to a laboratory that will conduct the sampling, do the analysis, and complete the annual submission forms for the firm. An estimated level of effort for evaluating alternative response to the surcharge is estimated at approximately two person-months over the past two years for fairly senior-level staff.

It seems clear that this firm's response to the sewer surcharge has been to minimize costs. Based on this decision rule, the firm continues to pay the surcharge rather than institute pretreatment methods at its plant. However, if the surcharge were to increase significantly in either relative or absolute terms, the firm's position would likely change. In this particular instance, the objective of the surcharge seems to have been met. The firm is paying a more proportionate share of the cost of treating its industrial waste, and it has not been faced with taking on the task of wastewater management.

11 Fruit and Vegetable Canning

Introduction to the Plant

This western company produces a variety of canned and frozen vegetables and fruits. Fruit commodities are a minor component of the total production. The firm is a cooperatively owned enterprise with an elected board of directors. The co-op membership is made up of approximately 300 growers. Each member has one vote in electing the board. Profits are shared by all growers and apportioned according to each grower's contribution to the total production.

There are three processing plants and one warehouse within the co-op.[1] All plants ship to the warehouse for distribution. Products are marketed under both the firm's label and various buyers' labels. The majority of the products enter a regional market. However, nationwide distribution also occurs, primarily under buyers' labels.

The plant under study was established in the 1930s and has experienced a number of expansions since. In 1978, 51,000 tons of raw product were processed. Most of the plant's production occurs from July through October. The plant staff consists of 100 permanent employees and approximately 1,000 to 1,100 seasonal employees during peak production.

Production Characteristics

The plant processes the following commodities: beets, carrots, corn, green beans, onions, squash, purple plums, strawberries, and cane berries. All products are both canned and frozen except purple plums, onions, and beets, which are canned only, and squash, which is frozen only. Of the total finished products, two-thirds are canned and one-third are frozen.

Growth over approximately the last five years has been modest. Output of canned products has been stable. Growth has, in large part, been due to increased production of frozen products. At present, approximately 51,000 tons of raw products are processed yearly.

Production is highly seasonal, with 90 percent of the annual production taking place from July through October. The seasonal behavior of the plant is determined by the growth and harvesting of fresh crops. Products must be processed as soon after harvest as possible to ensure product quality and freshness. Processing dates for each product vary from year to year in response to meteorological and climatological conditions. In some cases, production occurs at the rate of three shifts per day, seven days per week, during peak harvesting periods.

Labor is an important component of the plant operations. As many as fifty employees are required to run one production line per shift. Their involvement is largely in the areas of product-handling and inspection. Most of the people directly involved in production are employed on a seasonal basis. Many are local residents who return yearly. The permanent staff conducts the management, engineering, mechanical, and clerical functions in the plant. During the production season many of the permanent staff members supervise production-line operations.

Production Methods

Five major production lines are employed in the plant: for green beans, carrots and beets, corn, squash, and fruits. A schematic of each of these lines is presented in figures 11-1 through 11-5. Each line has a physically distinct set of production equipment. All fruit commodities are processed in basically the same series of operations. Therefore, only a generic discussion of the process flow will be presented. Processing of onions accounts for a small portion of the plant's total production and thus is not included in the discussions to follow.

Green Beans (Figure 11-1)

After removal from the storage bins, initial cleaning of the beans involves an air-cleaning technique in which loose dirt, rocks, leaves, and vines are removed. Residual materials obtained in the dry-cleaning operation can readily be handled dry. Wet-washing involves a combination of a flood-type washer, consisting of an immersion or soak tank, followed by a water-spray washer. Washwater is a major source of suspended solids.

Once the product has been cleaned, the sorting operation removes cull and other unusable whole units. In the trimming operation, the ends of the beans are snipped off, and blemished, bruised, or overripe segments are sorted out of the whole units. Many of the residuals from this operation are handled dry.

Depending on the product style, a cutting operation may be used. Grading involves separating the product by size and quality before further processing. To deactivate enzymes and to remove tissue gases, a hot-water, steam-blanching bath is used.

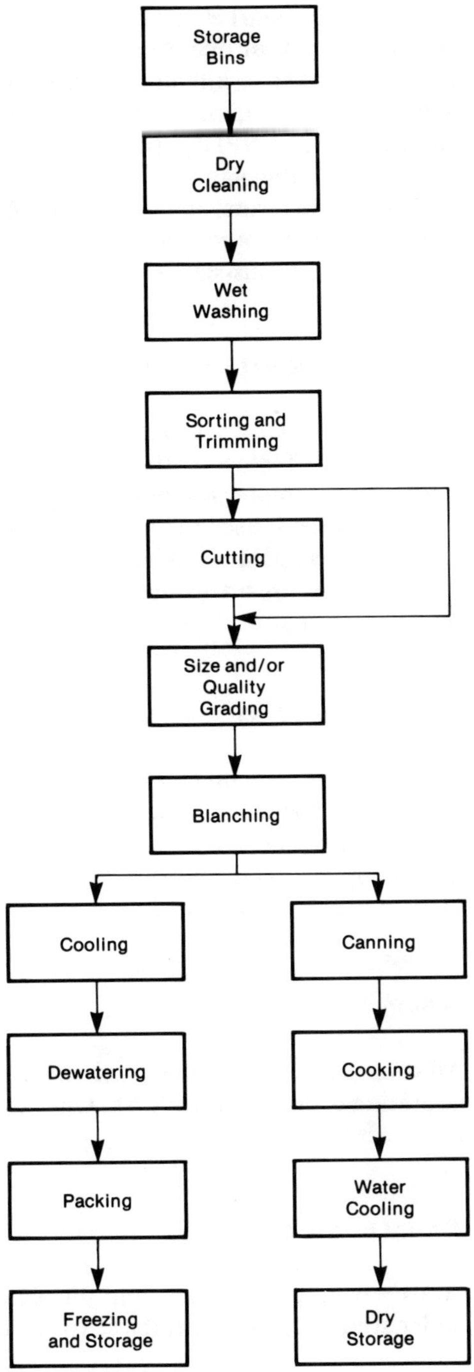

Figure 11-1. Process Flow—Green Beans

Beans are marketed as both a canned and frozen product. Both processes involve four operations. In the canning process, beans are packed in cans and cooked to kill all microorganisms. The cans are then water-cooled and stored for shipment. Presently, most of the can-cooling water is discharged under a NPDES permit to the storm sewer. A small portion is also reused within the plant.

To freeze beans, the product is cooled in a water bath to prevent the growth of microorganisms. Excess water is removed and the beans are packed into various-size containers, frozen, and stored for shipment.

Carrots and Beets (*Figure 11-2*)

Carrots and beets are processed in much the same way as beans.[2] The only significant difference is that the product must be peeled. To accomplish this, a hot, long-term lye-peeling process is used. After the lye-dunk, an abrasive peeler is used to remove the peel residuals. Much of the solid material from this operation is handled dry. This particular lye-peeling operation also performs the blanching operation that typically occurs just before canning or freezing.

Corn (*Figure 11-3*)

Processing of corn involves many of the same operations as previously discussed for green beans, except for husking of the corn. Husks are removed mechanically and handled as a dry residual. The corn is cut, blanched, graded by size and quality, and either canned or frozen. A steam-type blanching operation is used.

Squash (*Figure 11-4*)

All squash is frozen at the plant under study. The raw product is washed in a rotating cylinder with sprays. Once cleaned, the product is cut, blanched, and frozen.

Fruits (*Figure 11-5*)

Strawberries, cane berries, and purple plums are all processed at the plant under study. Unlike vegetables, fruits need no blanching. They are washed

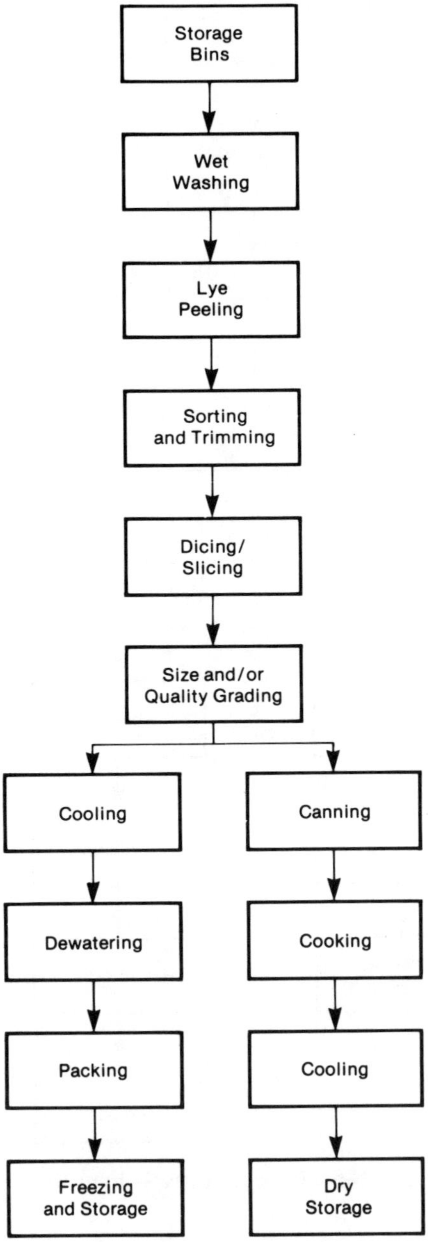

Figure 11-2. Process Flow—Carrots and Beets

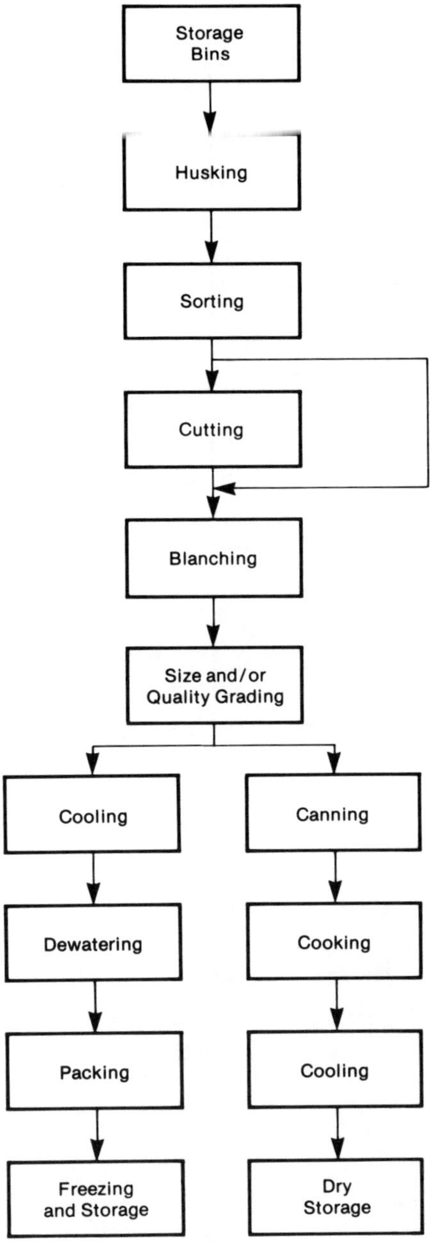

Figure 11-3. Process Flow—Corn

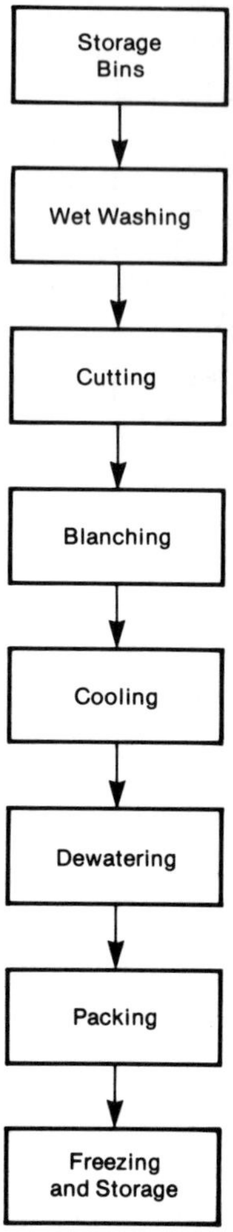

Figure 11-4. Process Flow—Squash

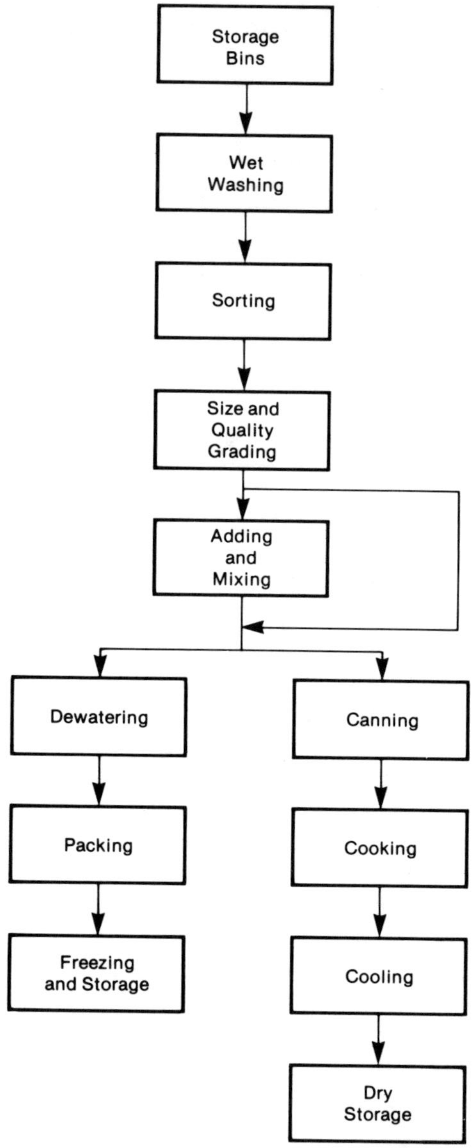

Figure 11-5. Process Flow—Fruits

in a spray-type washer, sorted to remove unusable whole units, and graded by size and quality. In the addition and mixing operation, various other constituents, such as syrups, juices, water, and the like, are added and mixed with the main product. Both canned and frozen fruits are produced.

Production Equipment

The production methods described for fruits and vegetables have been employed for a number of years without significant changes. However, upgrading of production equipment has occurred. For example, there has been a change in the variety of green beans processed at the plant from hand-picked pole beans to mechanically picked bush beans. Many more residual materials, such as dirt, leaves, and stems, are included with the bush beans, due to the mechanical picker. Washing equipment designed for the hand-picked pole beans was found to be inadequate. Pole-bean washers typically consisted of some arrangement of spray-type washers. These could not effectively remove the residuals from the bush beans and thus were replaced, usually with flood-type washers.

In deciding on equipment changes, all of the following factors are considered: labor savings, increased production, energy savings, and water use and wastewater production. As described, the production processes are very labor-intensive. Therefore, labor savings is in many cases the most important factor in making equipment choices. Increasing production volume and speed is also an important consideration. Energy use and water use and wastewater production are accounted for in making equipment changes, but to a lesser extent. Further discussion of water use and wastewater production will be presented in the sections to follow.

Context for Response to Charges

Environmental regulations of concern to the plant relate only to its wastewater-disposal practices. There are no applicable air-pollution regulations. Solid residuals generated by the plant are hauled away by private contractors for disposal by land application or for use as animal feed.

The plant is charged for all discharges to the municipal system. The city sharply increased its sewer rates 1 July 1976, when its expanded wastewater treatment plant went into operation. Industrial rates for flow, BOD, and TSS discharges were increased about 200 percent.

The new industrial charge formula consists of a constant monthly demand charge and a loading or variable charge. The demand charge depends on the maximum monthly utilization of the municipal system in the previous year, while the loading charge depends on the system usage in the month in question. The main objectives are to cover costs and to make industry pay its fair share. The loading charge is designed to recover industry's share of operation and maintenance costs, while demand charges relate to capacity and capital costs. Because the canneries in the system exert a large peak-loading, charges that explicitly distinguish between capacity, capital costs, and operation and maintenance costs were felt to be most equitable.

Earlier (prior to July 1976), industrial charges were based entirely on recorded effluents in the month of maximum BOD demand in the preceding year and thus were more nearly capacity charges than loading charges. As such, the old system did not provide a strong incentive to reduce flow, BOD, or TSS, except in the month of maximum BOD demand.

Beginning in March 1977, a permit system was instituted. Permits are issued annually, and they place certain restrictions on the plant's use of the municipal system. Permit specifications include:

1. Limits on the maximum daily and monthly total flow, BOD, and TSS discharges.
2. Continuous wastewater metering and daily twenty-four-hour, flow-proportional, composite samples of wastewater production.[3] The plant estimates its self-monitoring costs to be approximately $3,000 per year. Samples are collected and analyzed by the city daily.
3. Adjustment of pH levels in wastewater.

Flow, BOD, and TSS limits specified by the permit are based on the operating history of the plant. The intent of the discharge limits is to induce the plant to undertake pretreatment measures. Heretofore, the city has not enforced the permit limitations. Beginning in the 1979 season, the city intended to enforce permit limits on a cooperative basis, working with the plants to correct any problems that might occur. To date, the plant under consideration has not violated its permit limits.

A city ordinance also requires that all cannery wastes be passed through a 20-mesh screening operation and a gravity-settling system with a retention time of at least two hours. All canneries in the city have screening systems. The settling requirement has not been strictly enforced. However, the plant being studied does have a settling basin with the required two-hour retention time. The plant estimates that the cost of screening and settling is approximately $46,000 per year. This includes capital depreciation, labor, sludge disposal, truck expenses, and utilities.

During the 1976 and 1977 operating seasons, the area faced shortages in its water supply due to unusually low precipitation. The plant (and all plants in the city) was requested to voluntarily institute water-conservation measures. In complying with this request, the plant's wastewater flow was also significantly reduced. Although sewer-charge reduction was a beneficial result of the plant's water-conservation measures, it was only a secondary reason for the measures taken. The specific actions taken will be discussed in the next section.

The plant is well aware that the surcharges exist. Presently, the annual charge is $75,000 to $80,000. The technical director has primary responsibility for the environmental controls at all three plants in the co-op. The plant actively strives to minimize its charges. The cost effectiveness of providing various degrees of pretreatment as opposed to paying the charge is continuously being reexamined. The plant manager actively supports such activities.

Response to Charges

The following flow reduction measures have taken place over the last three to five years:

1. Beginning in 1975, a NPDES permit was obtained to discharge the can-cooling water to the storm sewer instead of to the sanitary sewer. The effluent is passed through a chlorination system before being discharged. Cost of the chlorination system was about $8,000. A small portion of the can-cooling water is diverted to the reuse system.
2. A personnel-training program was instituted in 1977 and is being carried out on an on-going basis. Foremen are responsible for training employees in the proper use of water, especially in cleanup activities. In each shift the plant water use is monitored and posted as an incentive to minimize water use.
3. A water-reuse system was installed in 1978. Final product wash water and blanch water, primarily from the green bean production line, is chlorinated and reused for initial washing. The chlorination system cost $13,000.
4. In 1978, shutoff valves were installed in the ends of all hoses.

Although potential sewer-charge savings were considered in the decision-making process, requests by the city that the plant reduce its water use due to drought conditions were the prime reason for the changes in plant operations. Based on somewhat uncertain figures, there was a net cost to the plant despite savings in water and sewer charges.

Two strength-reduction measures have taken place at the plant. In 1976, additional 20-mesh screening capacity was installed. Upgrading made it possible to remove solids more effectively during peak loads. Second, before 1977, the residuals from the lye peeling of beets and carrots were flushed into the plant's effluent stream. Beginning in 1977, the lye-peeler residuals were, for the most part, separated from the waste stream. They are now disposed of by land application. Savings in sewer charges were cited as the primary reason for the strength-reduction measures described.

Figure 11–6 presents the history of the plant's wastewater discharges to the municipal system. Yearly loadings of flow, BOD, and TSS are shown for the period 1965 through 1977. Since 1973, a steady decrease in flow can

Figure 11-6. Wastewater Discharge History—Fruit and Vegetable Canning

be observed, despite modest production increases. A dramatic decrease took place between 1976 and 1977. As discussed earlier, water-conservation measures in response to supply shortages were largely responsible for the reduction. BOD and TSS loadings vary significantly from year to year. Wastewater strength is highly dependent on production quantities and in particular on product mix. As mentioned earlier, this information was not available from the plant.

The cost effectiveness of various pretreatment methods is being reexamined on a continuing basis. Improved settling techniques being seriously considered include coagulation/flocculation and baffles in settling ponds. Biological treatment for BOD removal (BOD accounts for about 75 percent of plant's sewer charges) has been considered. In-house treatability and cost studies have been undertaken. BOD treatment has been rejected for a number of reasons, including startup and shutdown problems, variable waste load and make-up, and capacity requirements for treatment of peak flows. A good deal of concern was expressed for possible significant increases in sewer rates. Cost increases and restrictions set by the permit limitations are perceived as potential severe constraints on the plant operations.

Notes

1. Of the three plants, two process a mix of fruit and vegetables and the other processes green beans only.
2. In 1977 beet processing was shifted from another plant within the co-op to the plant being studied.
3. Self-monitoring was in effect prior to the start of the permit system.

12 Industrial Laundry

Introduction to the Plant

This is a large-scale, full-service industrial laundry, which concentrates on uniforms, mops, mats, and other industrial cloth uses, and is part of a national firm. The managers believe that they are cleaning up environmental problems (or at least concentrating them) and find that the strongest waste loadings come from big printers and similar industries.

The operation in the district employs about 160 people directly in one plant, in a twelve-hour, one-and-a-half-shift operation. They wash about 230,000 to 250,000 pounds and dry-clean 20,000 to 25,000 pounds each week. Dollar volume has grown about 45 percent in the last five years, with poundage up 15 percent or so in the last two years (and slower growth before then). The work is not seasonal, and the plant has been in operation for a long time.

Production, Operations, and Physical Plant

The wash operation appears similar to others visited, with the major steps of sorting, washing, water extraction, and ironing. No details were gathered on the actual water use per wash or the mix of detergents, but the goods washed were clearly more heavily soiled than, for example, the linens washed at several other sites. The interesting points about the production processes are changes made over the last several years, which are discussed later in this chapter.

Decision-Making and Response to Charges

The plant operators and the national engineering staff appear to have been active in trying to reduce waste loads and water consumption. Points that they mentioned include:

1. Regularly working with vendors on the best mix of detergent, machine time, and water use, with significant improvements (from 4.5 gallons per pound to about 3 to 3.5).

2. Installation about five years ago of a Washex dry cleaner, which is a closed system used primarily for mops and which recycles the oil used for cleaning through a vacuum filter to extract the dirt.
3. Installation one year ago of a new rug washer, which is supposed to recycle much of the water used, but which so far has been ineffective.
4. Installation of pretreatment, including shaker screens and an oil skimmer, before the water reaches the sewer.

The Washex was a $400,000 investment in 1974, and is expected to have a long payback time. Figure 12-1 shows the general flow for the Washex system. This system was installed on the recommendations of the corporate engineering staff and is used primarily for mops and mats, which are the dirtiest items. In essence, it is an oil-based dry cleaner, with recycling of the oil after dirt removal, so that solid waste is produced rather than water-

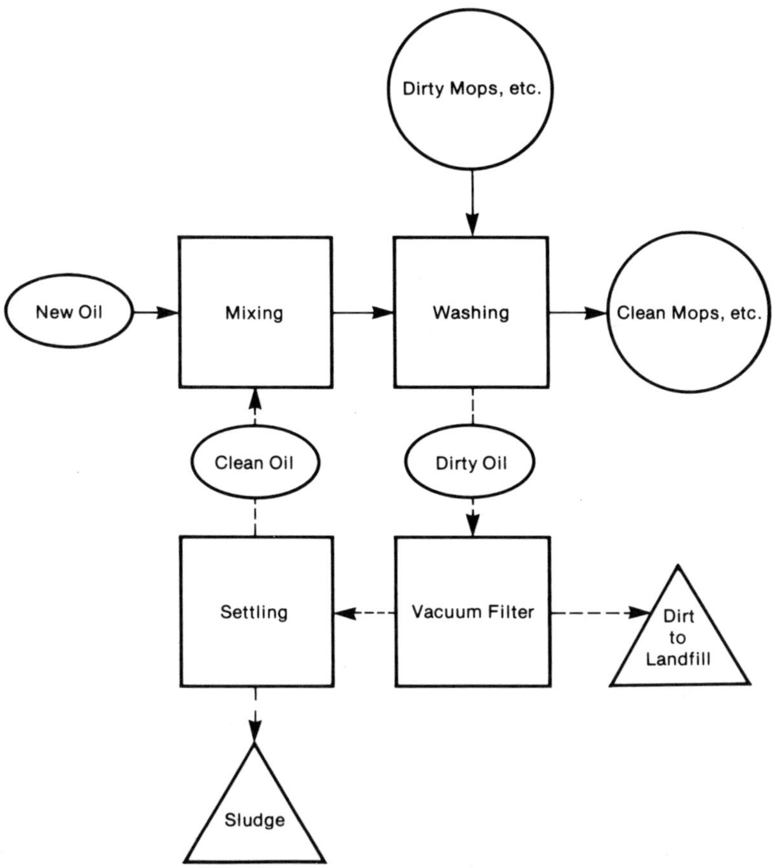

Figure 12-1. Process Flow—Mop Line

borne waste. We did not gather information on its relative cleaning efficiency, but it at least captures the residuals in a readily disposable form (the plant manager claims no problems in solid-waste disposal at present).

As mentioned earlier, wastewater pretreatment includes passing the wastes over a shaker screen, to remove threads and other large solids, and then into a settling chamber. The settled wastewater is then skimmed of oil and other floatables by a mechanical skimmer built at the plant. This system is claimed to work reasonably well, although the firm has tried at least one other method: disposable "pillows" that sit on the surface and soak up the oil. These were unsuccessful because the oil volume was too large.

The plant staff believed that that option would be more attractive than fancier pretreatment but clearly undesirable, with respect to both profits and general environmental goals. Chemical treatment was rated to cost about $250,000 for installation, based on other facilities, plus costs. The firm pays $7,000 a year to maintain the existing treatment system.

In general, then, they believed the charge system met their needs well and was less costly than separate treatment. However, they also believed that they might have to require better control over some of their customers (or refuse them altogether), under tougher standards, for substances like lead, or if they were faced with a regulatory limit on BOD or SS. As it is, they have charged some customers more for dirtier laundry and have convinced their customers to change some of their housekeeping practices. For example, one printer was mentioned as giving back rags and mops saturated with ink.

Table 12-1 presents the historical record, which clearly shows a major drop in wastewater concentration between 1974 and 1975, apparently

Table 12-1
Wastewater Discharge History, Industrial Laundry

	1974	*1975*	*1976*	*1977*	*1978*	*1979*[a]
Production[b]	.85	.87	.91	.93	1.0	1.10
Surcharge volume (mg)[c]	34.5	47.0	47.4	51.3	43.4	37.9
Surcharge (thousand dollars)	6.4	6.3	7.9	14.8	26.7	15.0
BOD (ppm)	2,166	1,450	1,411	1,294	1,697	1,901
TSS (ppm)	1,870	1,124	1,084	916	1,731	1,121
BOD/unit production	1.19	1.06	1.00	.97	1.0	.89
TSS/unit production	1.01	.81	.75	.67	1.0	.52
Flow/unit production	.94	1.24	1.20	1.27	1.0	.79

[a] Projected on limited data.
[b] Standardized to 1978.
[c] Million gallons; 92 percent of metered water from 1/78 on.

related to the investment in the Washex for the mop business as well as the shaker and skimmer. There also has been a recent upturn, apparently related to the construction of a heat-exchange unit, which has reduced use of the screens and oil skimmer for the last several months, and to one extremely high measurement in January 1978 (which may not be typical and which they are contesting).

Two points are worth repeating about this facility: the effectiveness of the closed-system investment for some of the dirtiest items, particularly the mops; and the possibility of a partial shutdown (refusing some dirty business) under tighter regulations on waste constituents, such as lead.

Note

1. It was sold to corporate headquarters as having a five-year payback period. In fact, plant management believes it will take at least 10 years.

13 Meat Packing

Introduction to the Plant

This family-owned plant is both a "baby beef" packing house and a producer of sausages and packaged luncheon meats. The younger-generation family members, recently out of school, have taken over some of the managerial responsibilities. In 1974, when it was first constructed, the plant was half the current size and was only a small slaughterhouse. The sausage kitchen was added in 1955, and in 1965 a new kill floor was installed. In the mid-1960s they stopped doing their own rendering and hide-curing and started sending the materials out to nearby facilities.

The plant processes 1,500 to 2,000 head per week of yearling animals and produces about 100,000 pounds per week of sausage products and packaged luncheon meats. The sausage production has remained constant over the past five years, but the slaughtering is down 25 percent. The dollar value of production has remained the same, however. It is still bigger than most meat packers and sausage producers, which employ, on average, 64 and 44 people, respectively. This plant employs 130 people in what is essentially one staggered shift. Different activities begin at different times.

The production is somewhat seasonal—around Christmas and New Year is their lowest point. In the summer, sausage production is increased.

Production, Operations, and Physical Plant

The sausage kitchen and slaughtering areas are separate, but both put wastewater into a common system. Figures 13-1 and 13-2 represent their processes. The animals are brought in by truck and held overnight in a pen. The manure from the pens is washed into the plant's sewers. The animals are brought to the kill floor, killed, and the blood is drained to a centrifuging unit owned and maintained by a blood-processing company. The hides are removed and stored outside in a truck (the plant does no dehairing) for daily pickup by a processor. After evisceration, the carcass is cut into sides and stored shrouded overnight in a chill room. The carcasses are then unshrouded and moved to a holding cooler. Sawdust is spread over the floor of the holding area to absorb any drippings and to make the area safer to walk on. The floor is cleaned and the sawdust is changed daily.

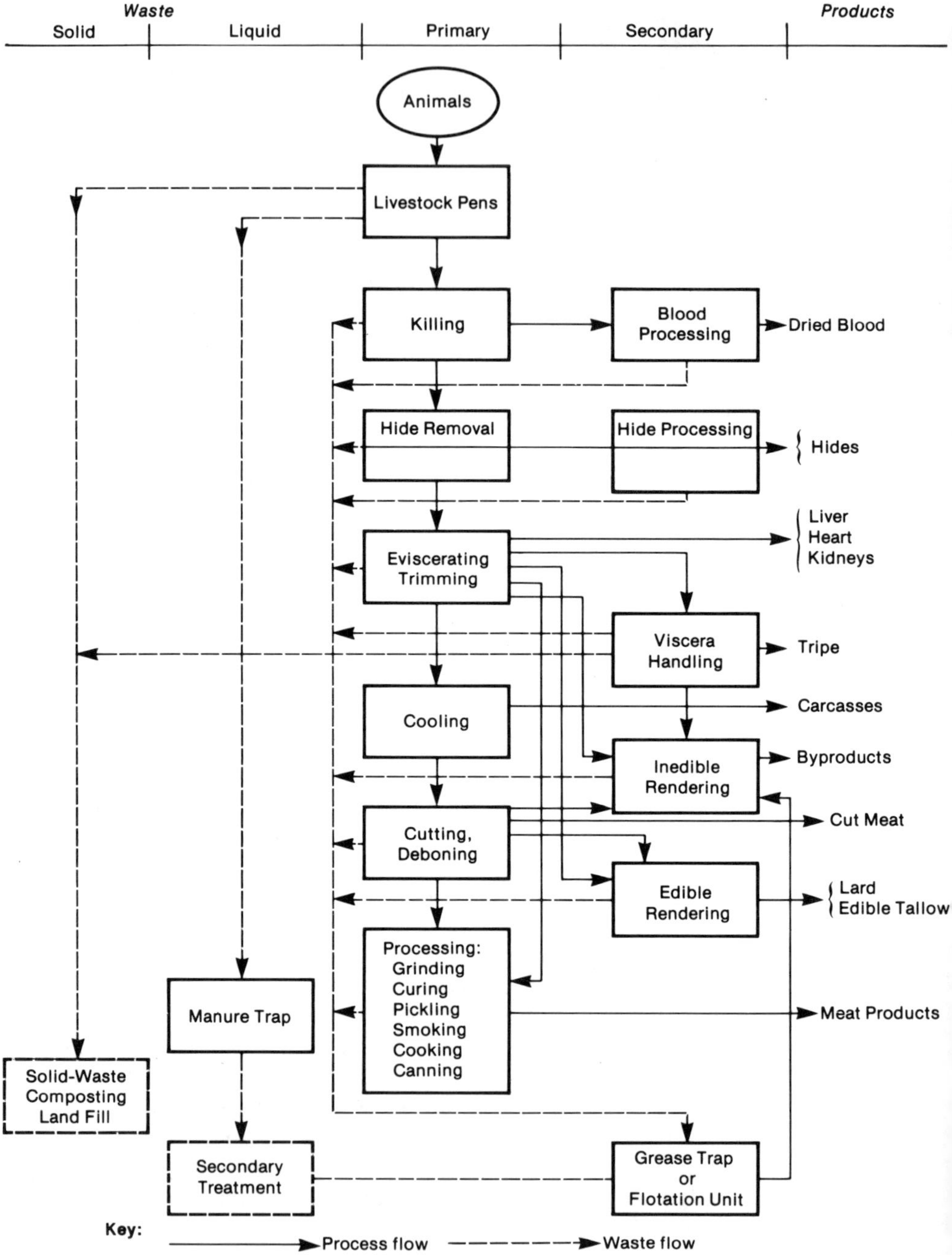

Figure 13-1. Process Flow—Meat Packing

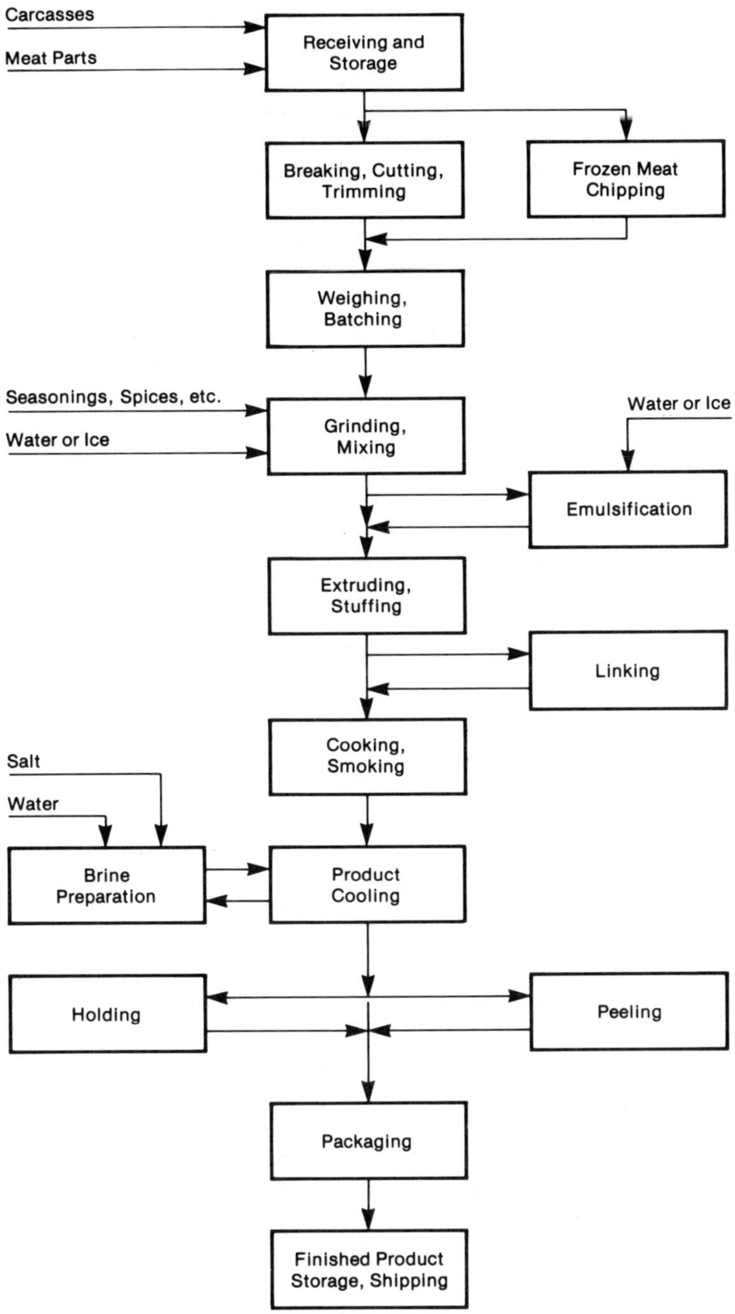

Figure 13–2. Process Flow—Sausages

The viscera are cleaned in a separate area. The paunch manure goes to a separator, to remove most of the liquid, and then to a landfill. The company attempted to find a market for the manure, but with no success. One of the problems was storage. The viscera themselves go to a renderer.

Most of the wastewater, then, comes from the frequent washdowns of the kill floor and the packing areas. In addition to being cleaned periodically throughout the day, each area has a separate clean-up crew, which works at the end of each shift. They clean and sanitize with 180°F water.

The sausage kitchen uses frozen meat and some materials from the packing house. The frozen material is tempered in water and then chipped, ground, weighed, spiced, chopped, emulsified, and stuffed. After stuffing, the product is showered, smoked, and showered once again. The casing is removed when appropriate, and then the product is packaged.

Most of the wastewater comes from the equipment washdowns at the end of each day and from the showering. The showering water goes to a holding tank and is subsequently used to wash down the barns. The wastewater from the slaughterhouse, sausage kitchen, and barn washdowns, and from the manure separator, flows to an Imhoff tank. From there it passes through a grease trap and out into the city's sewer lines. The tank and manure screen requires one full-time operator, who skims the grease trap regularly throughout the day and who will remove some sludge daily. The sludge goes to the sanitary landfill, the grease to the renderer.

Decision-Making and Response to Charges

Table 13-1 describes production, wastewater, and charges over time. The plant has had both the manure separator and the Imhoff tank for years. They were originally installed when the company did its own rendering. There is little else the company would consider doing; paying the surcharge is less expensive than installing a more extensive treatment system. One possibility considered by the company is adding a commercial bacteria enzyme product to the settling tank, which will supposedly break down the grease and help settle the solids, but as yet they have taken no action in this area.

If they were building a new facility, they might consider adding some lagoons—they have seen them successfully operated at other plants.

In response to their preference for quotas or market incentives, their response was, "The last thing we need around here are more regulations." The company would prefer to pay the surcharges for the services performed by the city's waste-treatment plants rather than be regulated as to the amount of waste they could discharge. They are in favor of the surcharge because they feel it not only ensures that the wastes are being handled properly but it is also equitable: firms must pay for any wastes they produce.

Table 13-1
Wastewater Discharge History, Meat-Packing Plant

	FY 1974	*FY 1975*	*FY 1976*	*FY 1977*	*FY 1978*	*FY 1978*[a]
Production[b]	1.24	1.19	1.12	1.07	1	.93
Flow (million gallons)	22.9	27.6	31.9	26.2	26.2	18.3
Surcharge (thousand dollars)	2.4	3.8	3.7	7.8	8.9	4.7
BOD (ppm)	1,312	1,402	1,068	1,588	1,000	1,095
SS (ppm)	1,187	1,236	1,273	810	595	679
BOD/unit production[b]	.58	.78	.73	.93	1	.51
SS/unit production[b]	1.41	1.84	2.33	1.27	1	.86
Flow/unit production[b]	.70	.89	1.09	.93	1	.75

[a] Extrapolation from current trends.
[b] Standardized to 1978.

14 Ice Cream Manufacturing

Introduction to the Plant

One of the plants surveyed is involved in the manufacture of liquid mixes for ice cream and the production of frozen ice cream products. The main response to sewer charges has been an ongoing argument that the city's sampling program does not provide a representative database for wastewater treatment charges.

The plant, with approximately thirty employees, is a branch facility of a family-owned business involved in the manufacture and distribution of dairy products. A second processing plant and three distribution facilities are also located in the area. The studied plant, SIC code 2026, was acquired in 1969.

Production Characteristics

Currently, about 1.5 million gallons of frozen ice cream and 2 million gallons of liquid ice cream mix are produced annually. The liquid mix was added to the product line in April 1977. This occurred shortly after the arrival of a new plant manager. As a result, business has increased significantly over the past two years, but it is anticipated that future growth will be slow. It is assumed that the business is profitable, as the possibility of additional plant acquisition was mentioned.

The plant's two products are manufactured year-round. From October to March, the demand for the company's ice cream products drops slightly. This produces less than a 10 percent decrease in wastewater loading.

The facility is normally operated five days per week. At the current staffing level, two employees start at midnight. Four additional workers start at 4:00 A.M., the rest between 6:00 A.M. and 8:00 A.M. The daily production cycle is completed by 7 P.M. and resumes at midnight.

Production Methods

The production of liquid mix and frozen ice cream products is virtually identical, up to the steps of flavoring and freezing. Figure 14–1 shows the

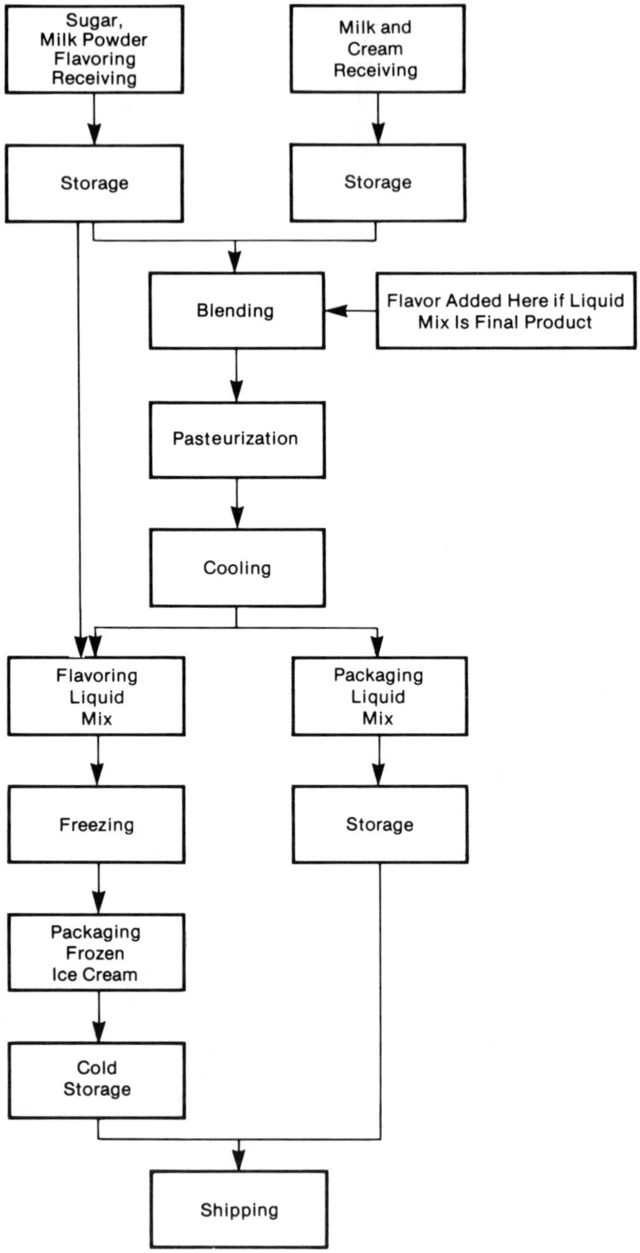

Figure 14-1. Process Flow—Ice Cream

basic production line for the two products. In a separate operation, condensed milk is also produced for use in the ice cream commodities.

The input materials include raw milk, condensed milk, cream, sugar, powdered milk, and flavorings. Milk and cream delivered to the plant are blended with the dry ingredients and then pasteurized in an unflavored state. After pasteurization, the liquid mix is packaged and stored for distribution. If hard ice cream is desired, the flavorings are then added. The flavored mixture is frozen, packed, and stored in cold storage for distribution.

In producing condensed milk, a standard process is used. The estimated production of condensed milk is currently 70,000 gallons per month. This is noteworthy because, in making this intermediate product, there is significant water loss discharged to the municipal sewers.

Wastewater from the plant for sanitary purposes is discharged into a sewer line separate from the discharge of production-related wastewaters. The major sources of discharges from production are washing and flushing operations.

Each piece of equipment used in producing liquid or frozen ice cream mix is cleaned at least once each working day. The equipment is cleaned in place, and each piece may be flushed separately from the rest of the line. Cleaning of some units more frequently than once per day can be necessitated by style changes. In some cases, however, a product style can be processed through the system without cleaning after manufacture of a previous style. If the complete line requires cleaning, equipment can be washed in series, running washwater through the system as if it were the product.

Wastewater is also produced in general housekeeping, such as cleaning of floors, machine exteriors, and so forth. Boiler washdown produces discharge. Finally, approximately two-thirds of the initial volume of raw milk to be condensed is lost in processing.

The intermediate product and raw materials lost in cleaning are major constituents of the wastewater. Also significant are the cleaning compounds and solvents. Table 14–1 summarizes the cleaning agents inventoried and used. All are used for cleaning of production units; the Clor-VL is used for boiler cleaning. Large amounts of water are used in flushing the cleaning solutions.

Context for Response to Charges

Environmental regulations of concern to the plant relate only to its wastewater-disposal practices. There are no applicable air-pollution regulations. No large amounts of solid residuals is generated. The plant is subject to state and federal regulation concerning ice cream production, designed to

Table 14-1
Cleaning Compounds

Stock Number	*Item*	*Chemical Description*	*Manufacturing Source*	*Solution Use, Concentration, and Quantity Stored*	*Comsumption Amount Time*	*pH*	*Disposal Frequency*
1	STU	Blend of sodium hydroxide, sodium metasilicate, sodium carbonate, sodium tripolyphosphate, nonionic wetting agent (biogradable)	Morgan-Gallacher	Concentration: .5% causticity in vacuum pan, .1% causticity in short-time unit. Reduced by rinse water in approximately 10:1 ratio. 1600 lbs stored.	1600 lbs/mo	13 12.7 9	Daily
2	Conc MSR	Mixture of phosphoric acid, water, nonionic wetting (biodegradable)	Morgan-Gallacher	1 gal Conc MSR to 200 gal water. 60 gal stored.	40 gal/mo	1	Daily
3	Clor VL	Blend of sodium hydroxide, sodium metasilicate, sodium carbonate, sodium tripolyphosphate, potassium dichloroisocyanurate, nonionic wetting agent	Morgan-Gallacher	1 oz Clor VL per 1 gal. water. 1200 lbs stored.	1200 lbs/mo	12.3	Daily
4	Tri Clor	Trisodium phosphate, sodium hypochlorite, sodium chloride	Morgan Gallacher	1.5 oz per 3.5 gal water. 300 lbs stored	300 lbs/mo	11.9	Daily
5	G-PCLOR	Sodium hypochlorite, water	Georgia Pacific Corporation	1 oz per 5 gal 30 gal stored.	60 gal/mo	7.5	Daily

protect public health and safety. These govern allowable additives, labeling, product attributes, and plant cleanliness.

The plant is charged for all discharges to the municipal system. The charges are based upon flow, COD, and TSS. Some form of industrial surcharge program has been in effect in the city since 1956. There have been numerous changes in the parameters monitored, the structure of the charge formula, and the rates of surcharge. Perhaps most significant was the 1975 elimination of BOD from the charge formula. The BOD measure was replaced by the COD parameter. This was based on the sewer district's belief that COD was a better measure of waste loadings, due to the interference with the accuracy of the BOD test caused by some types of industrial constituents. A second significant change was the elimination, in 1977, of minimum strengths above which charges apply.

Strength-monitoring by the city is performed at least quarterly, with a composite sample taken each time. The composite sample consists of grab samples taken four times per day at approximately two-hour intervals between 9:00 A.M. and 5:00 P.M. All testing for constituents is done by city-owned facilities. Annual strength-related charges are based on a simple average of observed concentrations from quarterly samples. Self-monitoring reports from certified laboratories can be used to supplement the city samples.

The flow volume is based on metered water-supply use. A determination can be made, based on a judgment by the superintendent of public works, of the amount of metered water that is lost in consumptive uses. Alternatively, the industrial plants have the option of installing effluent flow meters to directly measure, rather than estimate, wastewater discharge. Strength charges are calculated from average concentrations and the estimated discharge volume.

Beginning in 1975, a permit system was instituted under a source-control ordinance. Permits are issued annually and place certain restrictions on the plant's use of the municipal wastewater-collecting and treatment system. The ordinance limits permissible discharge concentrations of a variety of materials. Permit specifications include limits on the maximum daily flow as well as maximum COD and TSS concentrations and specification of a self-monitoring program of testing for COD, TSS, and a variety of restricted materials. The intent of the ordinance is to ensure that the city is able to meet its own effluent-quality requirements and to encourage pretreatment by industry in compliance with EPA's point-source guidelines. Restricted constituents include toxic materials, heavy metals, and a number of compounds that are incompatible with the municipal biological-treatment facilities.

This source-control program has created severe constraints in several industries. Although not the sort of economic incentive under study in this research, the source-control restrictions have motivated consideration, and in some cases implementation, of significant wastewater-management responses by local industries. The plant under study is in an industry that is not significantly affected by the source-control ordinance. In the process of trying to control restricted discharges, some permissible materials (for example, COD and TSS) are typically removed. When this has occurred, the resultant sewer-charge reductions are largely secondary to the primary goal of compliance with the source-control ordinance. Care is taken to try to discriminate among motivating factors in observed responses by the industrial plants under study.

During the period 1976–1977, much of the area experienced a significant drought. All of the plants in one study area receive water supply from a single utility. During much of 1977, that water utility imposed a target of 25-percent cutback in water consumption. This target reduction was passed through to the local industries as a voluntary water-conservation measure. In complying with this request, wastewater flow in many cases was significantly reduced. Although an associated sewer-charge reduction was achieved, it was not considered in all cases or was a secondary factor to the primary goal of reduced water consumption.

Response to Charges

No significant steps have been taken to reduce either the flow or the strength of wastewater. There has been a major effort to try to establish an accurate basis for wastewater-disposal charges. This has included two key steps. The first was a self-initiated sampling program to measure strengths of discharge. This program was undertaken to investigate possible deficiencies in the city's sampling program. The second step was an initial estimation of the consumptive use of water in the plant to reduce the charges that are presently based on metered water-supply use. These steps have been taken only in the last two years. In addition, informal consideration has been given to the possibility of flow equalization or pretreatment.

In order to resolve questions related to the city's industrial sampling program, Metcalf and Eddy, Inc., has been contracted to take twenty-four-hour composite samples of the plant's wastewater. The reasoning is that, due to the nature of the wastewater generating activities (batch discharges of washwaters resulting in large variations in effluent quality are likely), sporadic grab samples taken by the city would not capture these variations.

Self-monitoring activities were begun in January 1979. Table 14–2 illustrates the difference in estimated waste loadings using the city's 1978 samples and the first two composite samples taken by Metcalf and Eddy, Inc. Using either of the composite estimates, the city's 1978 calendar-year esti-

Table 14–2
Self-Monitored versus District Waste-Load Estimates

	COD (lbs/year)	*TSS (lbs/year)*
City, 1978	515,000	178,000
Sample, 1/24/79	105,000	15,000
Sample, 1/29/79	210,000	29,000

mates are roughly double those estimated on the basis of self-monitoring. The plant is planning to commission additional self-monitoring studies on a monthly basis.

An important step, not as yet resolved, is the ongoing attempt by the plant to arrange a credit with the sewer district for consumptive use within the plant. This attempt was initiated in October 1977. The metered water intake averages 90,000 gallon per day. Based on initial estimates, only 36,000 gallons per day of discharge have been documented, exclusive of effluent from milk condensation. The estimated effluent from milk condensation is roughly an average of 7,000 gallons per working day. Possible significant consumptive use of intake water occurs due to boiler make-up, evaporative cooling, and refrigeration activities. More detailed analysis of consumptive use by the plant is anticipated, but this could not be confirmed in interviews.

It is important to recognize that some industries, simply in product processing, may generate wastewater discharges significantly in excess of metered water intake. Care should be taken in the development of sewer-charge ordinances to account for this possibility.

Brief consideration has been given to the possibility of installing a holding tank to accomplish flow equalization. Pretreatment possibilities were mentioned but have not been given serious consideration.

Figure 14–2 illustrates the time history of COD and TSS concentrations in milligrams per liter. There appears to be a significant increase in strength. Water consumption over the period has moderately increased. Both the city and the plant are awaiting additional samples before drawing any conclusions from the time series.

In summary, the overall response has been to try to accurately estimate the basis for charges. The plant indicated a willingness to pay its fair share. The charge formula was not disputed. However, it was felt that the sampling program was not providing a representative database. Total charges for 1978, roughly $38,000, are estimated to be a relatively minor component of the costs of manufacture, and this is substantiated by the relatively low expenditures in trying to reduce charges.

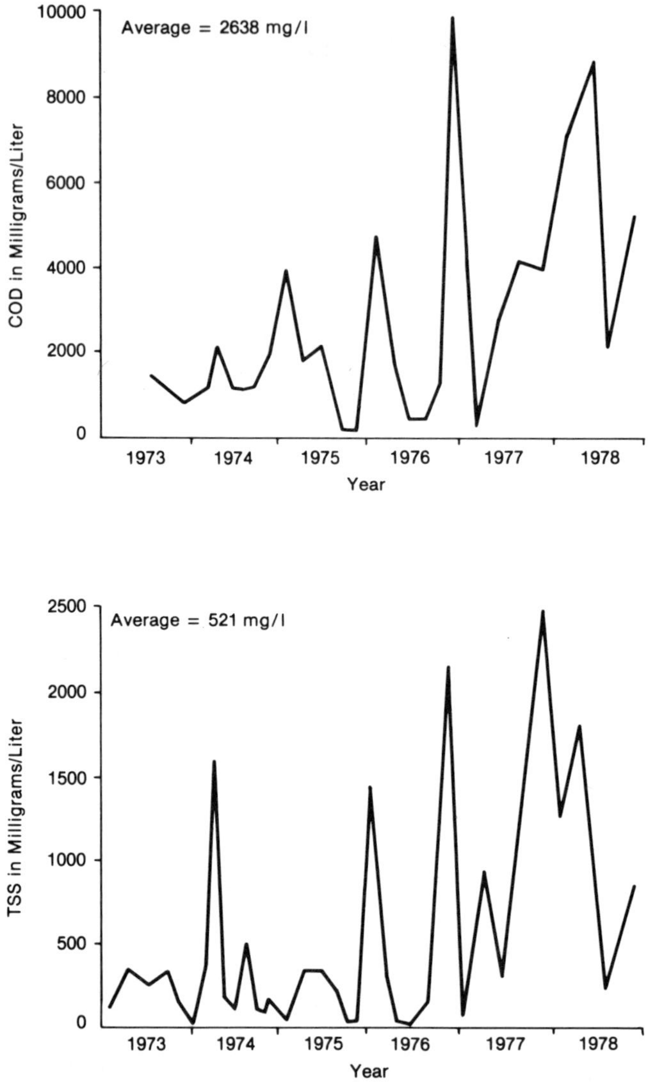

Figure 14-2. COD and TSS Histories—Ice Cream

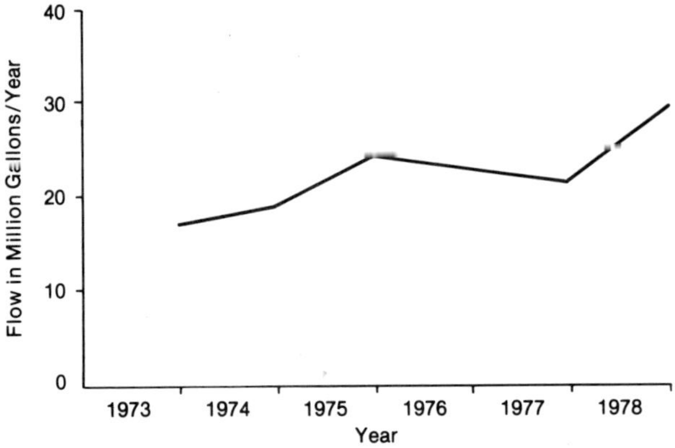

Figure 14-3. Flow History—Ice Cream

15 Poultry Processing

Introduction to the Plant

One of the western plants surveyed is involved in the processing of live turkeys. The firm is a cooperatively owned enterprise run by an elected board of directors. The co-op membership is made up of approximately forty-five growers. Each member has one vote in electing the board. Profits are shared by all growers and apportioned according to each member's contribution to the total production.

The co-op consists of a single processing plant and storage facility, both located in the city. Almost all the turkeys are raised in the region. The plant services a regional market. Both fresh and frozen turkeys are sold under the co-op's own label.

The cooperative was established in 1928. The main processing plant was built in 1956 and has undergone a number of expansions since. Presently, major production changes are being implemented at a cost of $500,000. In 1978, 1.5 million turkeys were processed. Most of the plant's production occurs from June through December. The plant staff consists of thirty permanent, full-time employees and approximately 225 seasonal employees during peak production.

Production Characteristics

The plant is currently (1978) processing 1.5 million turkeys per year. This corresponds to a total dressed weight (output from the plant) of 22 million pounds. There has been a decline in production since the plant's peak year of 1970, when 1.7 million birds were processed. The lowest output since then was in 1976, when 1.3 million birds were processed.

The trend experienced by this plant was even more severe on the northwest turkey industry in general. In 1970, 2.25 million birds were processed in the northwest by four plants. This last year (1978) the plant being studied accounted for the total regional production of 1.5 million birds. The other three plants that were processing in 1970 have since terminated operations.

It should be noted that the dressed weight of the turkeys can vary substantially from year to year. This is due in part to variations in the average bird weight. However another important factor is the prevalence of diseased birds. During the production process, all birds are examined by federal inspectors, and unwholesome carcasses are removed, thus lowering the average dressed weight. The presence of unwholesome birds also has had a significant impact on the plant's wastewater characteristics, as will be discussed in the sections to follow.

Production is highly seasonal, with approximately 80 percent of the annual production taking place in the six-month period beginning in June. The seasonal behavior of the plant is determined by the ability of the growers to raise turkeys throughout the year. During the winter months, the turkeys cannot be raised outside. Facilities to house the birds in a controlled climate must be used. Under such constraints, fewer birds are produced. In response to the supply of live turkeys, the plant operates at 100 percent of capacity between June and December and curtails its production to about 20 percent of capacity from January to June. Daily variations in output can also be significant, especially during the off-peak months, in response to the growers' delivery schedule.

Labor is an important component of the plant operations. Most of the people directly involved in production are employed on a seasonal basis. Many are local residents who return yearly. As many as 225 seasonal employees are involved in plant operations during peak output. During off-peak periods, approximately 100 people are employed on a part-time basis. A permanent, full-time staff of thirty employees performs the management, engineering, mechanical, and clerical functions of the plant. During peak production, about ten members of the permanent staff are involved in the direct supervision of production-line operations.

Production Methods

As noted earlier, the plant is currently making major alterations to its production line. Once completed, the new system will have a larger capacity and will be capable of completely processing birds in one working day. Previously, an overnight cooling operation was used. Faster market-delivery of the product will be possible. The new plant operations and layout were largely designed in-house, with some outside engineering consulting on an advisory basis. A description of the new production line is presented in this chapter. A schematic of the process flow is shown in figure 15-1. Processing takes place in a series of continuous operations.

Live turkeys are received from the growers on trucks and hung by their feet on a conveyor system for processing. Before leaving the plant, the trucks are cleaned to remove manure, feathers, and other materials. In the killing operation, the blood is drained from the birds. About 80 to 100 birds are in the unit at any one time. Presently, the blood is removed every four

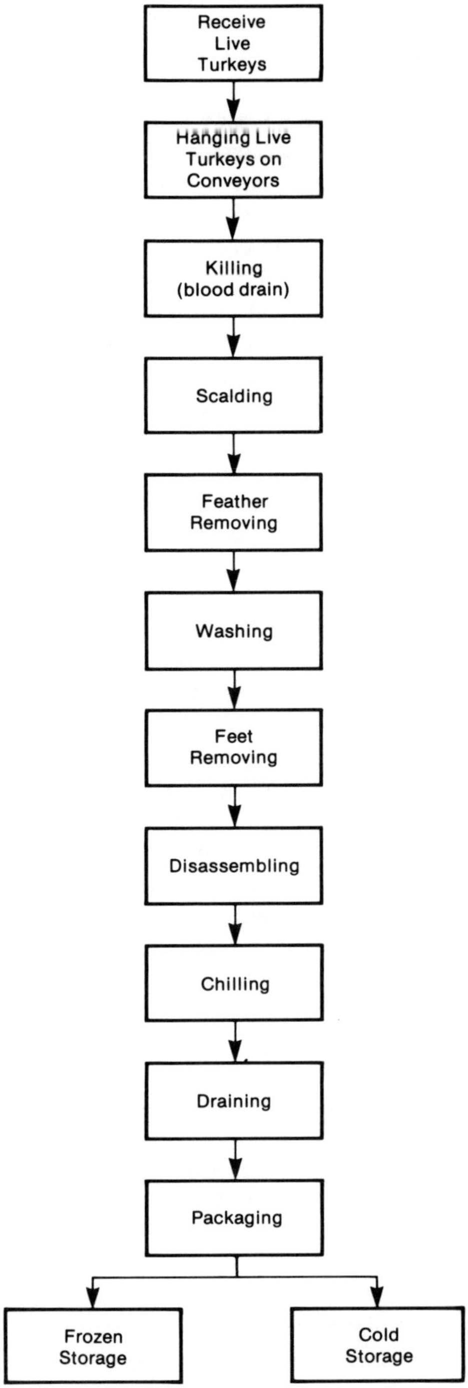

Figure 15-1. Process Flow—Turkey Preparation

hours. Most of it is handled as a solid, with only the final washwater entering the plant's main effluent stream. Blood washwater is characterized by high levels of BOD.

The birds are scalded for easier feather removal as well as for sanitation purposes. The feathers are then removed and flushed to the main effluent stream, using a system with a recirculating water supply. Any remaining feathers are removed in a spray-type washing operation. The birds are then removed from the conveyor, and the feet are removed and handled as a solid byproduct.

The disassembling operation essentially involves the removal of the inedible parts of the product; it is a labor-intensive procedure. Presently, unwholesome carcasses are removed and handled as a solid residual. Inedible portions are flushed to the main effluent stream. A large amount of washwater is used continuously in this operation.

To prevent the growth of microorganisms, a continuous, auger-type cooling system is used. The birds have a three-hour retention time in the chilling operation. Before packaging, excess water is drained from the product. The final commodity may be stored frozen of fresh.

About 60 percent of the total production staff is required to run all the operations before chilling. The daily eight-hour shift for these operations is scheduled to begin at 7:00 A.M. Three hours later, the first product has been processed through the cooling step, and the remaining operations are scheduled to begin at 10:00 A.M. A cleanup crew completely washes down and sanitizes the plant each night.

In deciding on the design and layout of the production line described, all of the following factors can be considered: labor savings, increased production, energy requirements, and water-use/wastewater production. For this plant, increased production was cited as the most important factor in making equipment choices. Water-use/wastewater production was also an important component in the decision-making process. Further discussion of water-use/wastewater production will be presented in the sections to follow.

Context for Response to Charges

Environmental regulations of concern to the plant relate only to its wastewater-disposal practices. There are no applicable air-pollution regulations. Solid residuals generated by the plant are sold as byproducts or hauled away by private contractors for land disposal. The plant is subject to state and federal regulations concerning food processing, designed to protect public health and safety. These govern allowable additives, labeling, product attributes, and plant cleanliness.

The plant's operations are under constant inspection by the U.S. Department of Agriculture (USDA). All production procedures are tightly controlled by USDA regulations. The design and layout of the plant's present production line was subject to government approval. The plant manager expressed the opinion that many regulations are outdated and result in increased wastewater volume and strength.

The plant is charged for all discharges to the municipal collection and treatment system. The city sharply increased its sewer rates 1 June 1976, when the expanded wastewater treatment plant went into operation. Industrial rates for flow, BOD, and TSS discharges were increased about 200 percent.

The new industrial charge formula consists of a constant monthly demand charge and a loading or variable charge. The demand charge depends on the maximum monthly utilization of the municipal system in the previous year, while the loading charge depends on the system usage in the month in question. The main objectives are to cover costs and make industry pay its fair share. The loading charge is designed to recover industry's share of operation and maintenance costs, while demand charges relate to capacity and capital costs.

Because canneries comprise a major portion of the industrial sector and exert a large peak loading on the system, charges that explicitly distinguish between capacity, capital costs, and operation and maintenance costs were believed to be most equitable. Wastewater from the plant is not as highly seasonal in flow or strength as cannery effluents.

Earlier (prior to July 1976), industrial charges were based entirely on recorded effluents in the month of maximum BOD demand in the preceding year and thus were more nearly capacity charges than loading charges. As such, the old system did not provide a strong incentive to reduce flow, BOD, or TSS, except in the month of maximum BOD demand.

Beginning in March 1977, a permit system was instituted. Permits are issued annually and place certain restrictions on the plant's use of the municipal system. Permit specifications include:

1. Limits on the maximum daily and monthly total flow, BOD, and TSS discharges.
2. Continuous wastewater metering and daily twenty-four hour, flow proportional, composite samples of wastewater production.[1] The plant estimated its self-monitoring costs to be approximately $10,000 per year, which is considerably higher than estimates by other plants subject to the same regulations. Samples are collected and analyzed by the city daily.
3. Adjustment of pH levels of wastewater.

Flow, BOD, and TSS limits specified by the permit are based on the operating history of the plant. The intent of the discharge limits is to encourage pretreatment by industry in compliance with EPA's point-sources guidelines. Heretofore, the city has not enforced the permit limitations. Beginning in the 1979 season, the city intended to enforce permit limits on a cooperative basis, working with the plants to correct any problems that might occur. To date, the plant under consideration has not violated its permit limits.

City ordinances also require that all industrial wastes be passed through a 20-mesh screening operation and a gravity settling system with a retention time of at least two hours. The plant does have a 40-mesh screening system. The settling requirement has not been strictly enforced, and the plant has not been required (or chosen) to install a settling system with the required two-hour retention time. Effluent from the plant is settled in a system with a residence time on the order of a few minutes. The plant estimates that the cost of pretreatment is $20,000 per year.

During the 1976 and 1977 operating seasons, the area faced shortages in water supply due to unusually low precipitation. All plants in the city were requested to voluntarily institute water-conservation measures. In complying with this request, wastewater flow in many cases was also significantly reduced. Although sewer-charge reduction was a beneficial result of the plant's water-conservation measures, it was only a secondary reason for the measures taken. The specific actions taken will be discussed in the section to follow.

Response to Charges

The plant is well aware that the charge program exists and has indicated a willingness to pay its fair share. The charge formula was not disputed. However, the plant actively strives to minimize its charges. Water-use/wastewater production was an important factor in the design and layout of the plant's current production-line overhaul. It should be noted that increased plant capacity was the primary goal of the changes. Also, all procedures were subject to USDA approval. The production-line changes that affect the plant's wastewater characteristics are described here.

Changes in the operating procedures of the killing unit have resulted in decreased water use and in decreased BOD loadings particularly. Originally, regulations required a continuous flow of washwater in the blood-draining area on both floor and walls. About six to seven years ago, approval was obtained to allow accumulation of blood on the floor for a maximum of four hours; and about two years ago, approval was obtained to allow accumulation of blood on the walls for a maximum of four hours. These changes made it possible to remove most of the blood residuals as a solid for a nonsewer disposal. The original water system is presently used for final cleaning of the unit.

In the feather-removing operation, a spray-type system with a recirculating water supply is used to flush the feathers from the unit. This water does not come in contact with the product. The feathers are removed from the effluent for disposal. Over time, a number of changes have been made in the equipment to separate out the feathers.

Major changes in the equipment, layout, and operating procedures of the disassembling step have affected both effluent flow and strength. As described previously, inedible portions of the product are separated from the edible parts. A large amount of water is used for hand and product rinsing. Approval has been received for the installation of timed shutoff valves on all rinse equipment, resulting in significant water savings. All solid residuals from the operation are conveyed in a fluming system, which reuses water from the chilling step.

Presently, unwholesome carcasses are removed from the production line as whole units. Regulations previously required the grinding of unwholesome carcasses for discharge to the main effluent stream. The grinding procedure significantly increased the BOD and TSS strength of the effluent. This is especially pronounced when the prevalence of disease substantially increases the number of unwholesome birds.

Product chilling is carried out in a continuous, auger-type unit. The product has a three-hour retention time in the system. The required overflow from the system is one gallon per turkey. The previous system was a batch-type unit in which the product was held for twenty-four hours. Ice water was used as the cooling media and required changing three to four times over the twenty-four-hour period. The new system requires half this water volume.

Beginning in 1968, water from the chilling operation was used for initial washdown of the turkey-delivery trucks and receiving area. City water is used for the final cleaning step.

Figure 15–2 is a schematic of the plant's effluent treatment scheme. Solid residuals, which consist largely of inedible turkey parts, are separated in a 40-mesh screening system. The liquid-effluent stream is reused to flush feathers from the feather-removing operation. An added benefit of the reuse system is the removal of additional solids that adhere to the feathers.

A time series of the plant's wastewater discharges to the municipal system is presented in figure 15–3. Yearly loadings of flow, BOD, and TSS are shown for the period 1965 through 1977. For BOD and TSS, there is a modest decreasing trend. In general, the discharge history is characterized by large variations in annual loadings as high as 50 to 100 percent. In most cases, there is a positive correlation between yearly flow, BOD, and TSS, which indicates that fluctuations in annual production volumes are responsible for the behavior. However, only modest variations in annual production, on the order of 10 to 20 percent, could be identified during the plant interview. The disposal of unwholesome whole turkeys due to the presence of disease was cited as an important factor in determining yearly loadings. However, this would only affect BOD and TSS loadings.

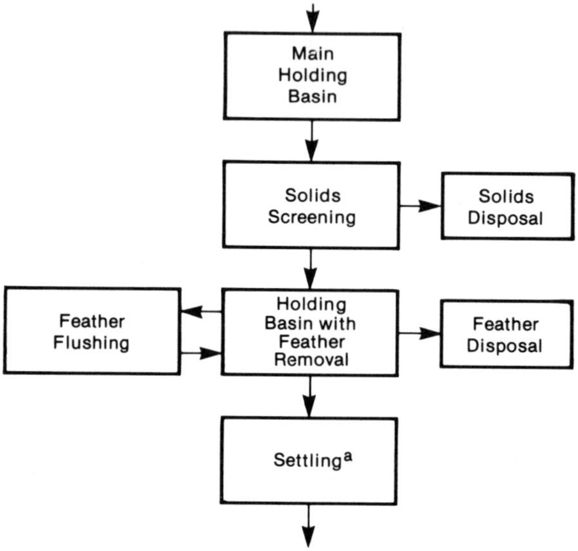

[a] Retention time on the order of a few minutes

Figure 15-2. Effluent Pretreatment System

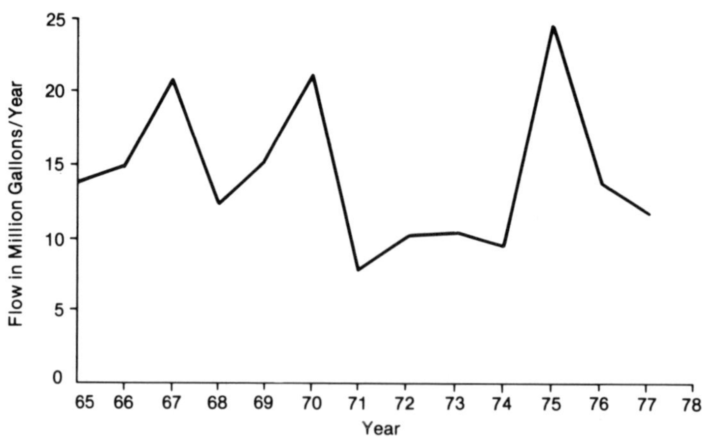

Figure 15-3. Wastewater Discharge History—Poultry Processing

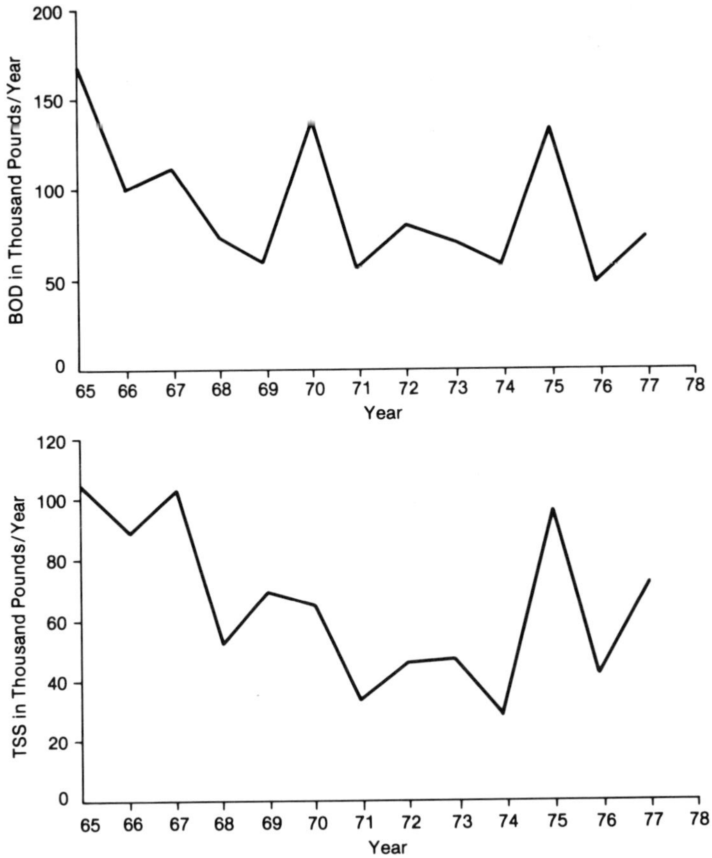

Figure 15-3 continued

During 1976 and 1977, there was a water-supply shortage in the area. For that period, the plant's BOD and TSS loading increased and flow volume decreased. Increased strength loadings can be attributed to production increases. Decreased flow indicates that an effective water-conservation program was initiated by the plant in response to the drought.

A number of alternative wastewater-control strategies have been considered. Most of the emphasis has been placed on in-plant control of wastewater. In the design of the production equipment, layout, and operating procedures, a conscious effort was made to minimize wastewater generation. However, this area is tightly controlled by USDA regulations. Some pretreatment alternatives, especially for BOD, which accounts for the largest portion of the plant's effluent charges, have been considered. Space limitations on the plant grounds were cited as an important constraint. In addition, the plant's effluent loading is variable over the year, making biologial treatment for BOD removal difficult.

Note

1. Self-monitoring was in effect prior to the start of the permit system.

16 Plating Job Shop

Introduction to the Plant

This plant is the only location of a privately owned heat-treating and plating firm. The firm is a job shop that does not do any fabrication. Work is performed largely for the automotive industry, especially providing small parts, such as seat-belt fasteners. The firm employs 65 persons in production and has annual sales of approximately $5 million. Although the firm has been at the same location for the last 30 years, most of the capital stock, estimated to be in excess of $60,000 per employee, or close to $4 million, has been purchased in the last ten years. Private bank financing has been used in these modernization efforts.

Production, Operations, and Physical Plant

The plant contains about 200,000 square feet of floor space and has two major areas of about equal size. The heat-treatment process, which is very energy-intensive and primarily uses natural gas, does not generate significant process-water-related wastewater flows. The electroplating process accounts for most of the wastewater generation.

The firm has invested heavily in labor-saving equipment, which has resulted in a high production/labor ratio. Currently, the plant is operated on a twenty-four-hour basis, with three shifts of about twenty persons each.

Inputs into the process are parts, energy, chemicals, and labor. The management estimates that energy costs have doubled over the past four years and that chemical costs have been rising at about 15 percent per year. Labor costs have also been increasing at about 7 to 10 percent per year.

A typical plating operation is utilized, with the parts carried through progressive plating tanks in baskets. Process water is currently collected in a centralized in-plant sewer, which delivers the wastewater to a pretreatment facility where cyanides are chlorinated and corrected for pH. Solids are separated from the wastewater by means of centrifuge, which leaves a 6 percent sludge. The effluent is then discharged to the district's sewers.

Decision-Making and Response to Charges

The surcharge has generally been considered too small a cost to warrant any pretreatment efforts. Table 16–1 summarizes the discharge history for the past five years. Given that the surcharge rates have more than doubled over the same period, these figures show that the actual amount of the wastewater flow has been relatively constant from this plant. The firm has been forced into pretreatment because of the district's limited-substances regulations. Surcharge impacts were relatively insignificant. There have been some reductions in BOD concentrations, largely as a result of process changes that had nothing to do with the implementation of the surcharge. Management views the biggest issue to be the excessive increases in surcharge rates (rates have approximately doubled in five years).

The installation of pretreatment facilities to meet limited-substances regulations has been quite costly. Currently, the firm has been faced with serious solid-waste-disposal constraints as a result of the district's crackdown on the disposal of industrial sludge. An independent waste hauler, who had been contracted to dispose of the pretreatment sludge, was found to be simply dumping the sludge down a manhole in another section of the district. The firm has not been able to identify an alternative hauler and is currently storing the sludge on-site.

Because of the strict enforcement of the limited-substance regulations and the problems with sludge disposal, the firm is currently engineering a new wastewater-management system. When the system is completed, wastewater from each of the five plating lines will be treated separately, with a maximum amount of chemical recovery and reuse. The management stressed, however, that the economic benefits of chemical recovery are insignificant relative to the additional costs. The management considers the entire effort to be an additional cost for pollution control, with no benefits. Effluents from the five separate pretreatment units will then be mixed, with a final pH adjustment before discharge to the district's system. It is unclear whether the installation of this elaborate pretreatment system will affect the

Table 16–1
Wastewater Discharge History, Plating Job Shop

Year	*1974*	*1975*	*1976*	*1977*	*1978*
Flow (million gal)	82	102	102	135	124
BOD (thousand pounds)	10	11	17	34	29
SS (thousand pounds)	46	69	69	80	51
Surcharge ($)	2,986	4,562	5,655	8,348	7,696

surcharge fees. However, the firm estimates it will cost $60,000 in additional treatment equipment, plus considerable increases in operational cost. The system should reduce solid-waste-management problems however, since the sludge will be suitable for landfilling.

The management estimates that 15 percent to 20 percent of the sales dollar is for water-pollution control, although less than 0.5 percent is spent on the surcharge program. The firm has been forced to raise prices to meet these costs and believes that it has been put in an unfair competitive position with the firms located in areas where pretreatment regulations are not strictly enforced.

17 Paint Manufacturing

Introduction to the Plant

One of the western plans surveyed is involved in the manufacture of paints and allied products (SIC 2852). The plant was established by the present owners in 1935. Substantial growth has taken place over time, and new commodities have been added to the product line.

The plant is part of a large corporation that produces a wide variety of chemical products and operates a number of plants nationwide. A total of twelve departments makes up the corporation. The studied facility is one of nine plants in the corporation's Fabrics and Finishes Department. Only product manufacturing (that is, no direct sales) takes place at the plant. Ultimately, products are sold under the corporate label to the industrial, consumer, refinishing, and automotive markets. The area serviced by the plant is generally west of the Rocky Mountains, with some nationwide distribution.

Production data were not made available in the plant interview. Seasonal variations in the product mix occur throughout the year in response to market demands. However, plant staffing is constant year-round and currently consists of 206 permanent, full-time employees.

Production Characteristics

The following commodities are manufactured: alkyd-base paint, acrylic-base paint, latex-base paint, epoxies, and thinners. Data giving total plant output and product mix were withheld as proprietary information. However, the plant did agree to release the information contained in the plant's application for a city-issued wastewater-discharge permit. Table 17-1 presents production data contained in the plant's permit application. Maximum and average outputs are given by market type and product. The products are not completely consistent with the list given, but it was the only available output measure. The volumes in table 17-1 were cited in the plant's 1976, 1977, and 1978 permit applications, which indicates that plant output has been relatively constant over the past few years.

The plant's total production rate is said to be moderately seasonal, in response to market demands. Product mix does change significantly

Table 17-1
Production Data

		Quantities (thousand gal/mo)	
Market Type	*Product Type(s)*	*Average*	*Maximum*
Automotive	Lacquers and enamel	60	100
Consumer	Emulsions and alkyds	39	60
Industrial	Enamels	28	37
Maintenance, finishing	Enamels, alkyds, and emulsions	10	11
Refinish	Alkyds	60	85

throughout the year, especially for automotive and house paints. For example, house-paint production typically increases by about 50 percent in April and May. Inventory-control measures are being taken to even out peak production loads. Table 17-2 shows the plant's monthly water-supply use for calendar year 1976. Monthly variations in volume are seen to be quite large, at about +30 percent of the mean. Water use is not necessarily strongly related to production quantities, but it does give some indication of the variability.

The plant operations are conducted throughout the year at a rate of five days per week, three shifts per day. The level of staffing is constant year-round and consists of 35 office and 171 production employees. Of the production employees, the distribution between the day, swing, and night shifts is 61, 23, and 16 percent, respectively. The production processes can vary widely from day to day.

Production Methods

The facility manufactures paints and allied products. Approximately 1,000 different raw materials are used, many of which are purchased from outside sources. Manufacturing activities vary according to market demand, and therefore a variety of materials can be used at any one time. The production line and key wastewater-generating operations are discussed in the following paragraphs.

The plant's two major products are oil-base and water-base paints. There is little difference in the production processes used. The major production difference is in the carrying agent: oil-base paints are dispersed in an oil mixture, while water-base paints are dispersed in water, with a biodegradable surfactant used as the dispersing agent. The next significant

Table 17-2
Monthly Metered Water Use, 1976

Month	*HCF*	*Percentage of Mean*
January	6,100	102
February	4,431	74
March	4,730	79
April	4,858	82
May	5,876	99
June	5,174	87
July	5,763	97
August	6,022	101
September	6,446	108
October	7,526	126
November	7,799	131
December	6,732	113
Total	71,457	
Mean	5,955	
Standard deviation	1,062	

difference is in the cleanup procedures. As the water-base paints contain surfactants, it is much easier to clean up the tubs with water. The tubs used to make oil-base paint are generally cleaned with an organic solvent, but cleaning with a strong caustic solution is also done.

Figure 17-1 presents a simple schematic of the process-flow for paint manufacture. All paints are made in batches. There are generally too many color formulations to make a continuous process feasible. There are three major steps in the paint manufacturing process:

1. Mixing and dispersing of raw materials, with grinding typically for oil-base paint.
2. Tinting and thinning.
3. Filling operations.

For oil-base paints, the major raw materials include pigments, oils, resins, tints, and thinners. The mixing and grinding of raw materials for oil-base paints are accomplished in one production step. For high-gloss paints, the pigments and a portion of the binder and vehicle are mixed into a paste of a specified consistency. This paste is fed to a grinder, which disperses the pigments by breaking down particle aggregates rather than by reducing the particle size. Two types of grinders are used for this purpose: pebble or

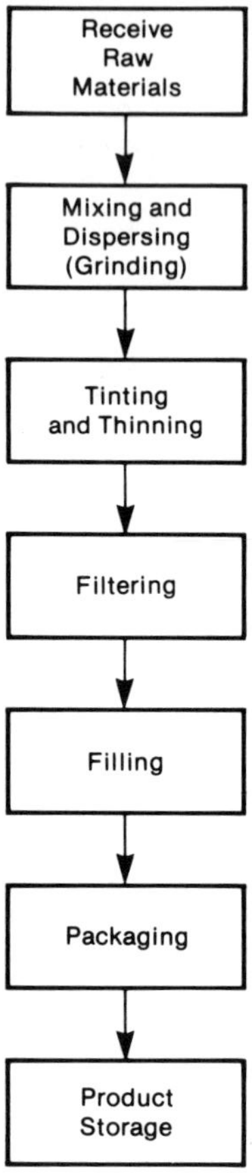

Figure 17-1. Process Flow—
Oil- and Water-Base Paints

steel-ball mills, or roll-type mills. Other paints are mixed and dispersed in the mixer using a saw-toothed dispersing blade.

In the next stage of production, the paint is transferred to tinting and thinning tanks. Here, the remaining binder and liquid, as well as various additives and tinting colors, are incorporated. The paint is then analyzed and the composition is adjusted as necessary to obtain the correct formulation for the type of paint being produced. The finished product is then transferred through a series of operations where it is filtered, filled into containers, packaged, and labeled. These operations are mechanized. Final products are stored on pallets.

The paint remaining on the sides of the tubs or tanks may be allowed to drain naturally and the "cleavage," as it is called, wasted, or the sides may be cleaned with a squeegee during the filling operation until only a small quantity of paint remains. The final cleanup of the tubs generally consists of flushing with an oil-base solvent until clean. The dirty solvent is redistilled for further use. Tanks are also cleaned with a strong caustic solution. The caustic is used to remove wastes that may have hardened in the tanks and would not be amenable to cleanup with solvent. Wastewater from the caustic wash is treated before discharge, as is all plant wastewater.

Raw materials for manufacturing of water-base paints include pigments, oils, surfactants, tints, and water as the carrying agent. Water-base paints are produced by a method slightly different from that for oil-base paints. The pigments and extending agents are usually received in proper particle size, and the dispersion of the pigment, surfactant, and binder into the vehicle is accomplished with a saw-toothed disperser. The paint is transferred to special tanks for final thinning and tinting. Once the formulation is correct, the paint is processed through the same series of operations as described for oil-base paints.

As in the oil-base paint operation, as much product as possible may be removed from the sides of the tub or tank before final cleanup starts. Cleanup of the water-base paint tubs is done simply by washing the sides with a garden hose or a more sophisticated washing device. The washwater is treated before discharge.

The plant also manufacturs the synthetic resins used—either the usual alkyd resins, a water-soluble alkyd resin, or an acrylic resin. The manufacrure of either type involves an esterification process in which polybasic acids and polyhydric alcohols react with various oils or fatty acids. The raw materials are fed into a large reactor (kettle) equipped with an agitator. The kettle is then heated to the specified reaction temperature. Most alkyd resins are manufactured at about 200°C (392°F). The heated resins are cooled, filtered, and stored for use in paint production or for sale.

A summary of the key wastewater-generating steps is presented in figure 17-2. Major sources originate in the main manufacturing plant, the cleanup facilities, and the resin-manufacturing plant. All plant effluent is treated before discharge to the municipal collection and treatment system. A description of the plant's pretreatment scheme is contained in the section on response to charges. The maximum and average daily wastewater flow is 8,544 and 6,144 gallons per day, respectively.

Context for Response to Charges

The plant is charged for all discharges to the municipal system. The charges are based upon flow, COD, and TSS. Some form of industrial surcharge program has been in effect in the district since 1956. There have been

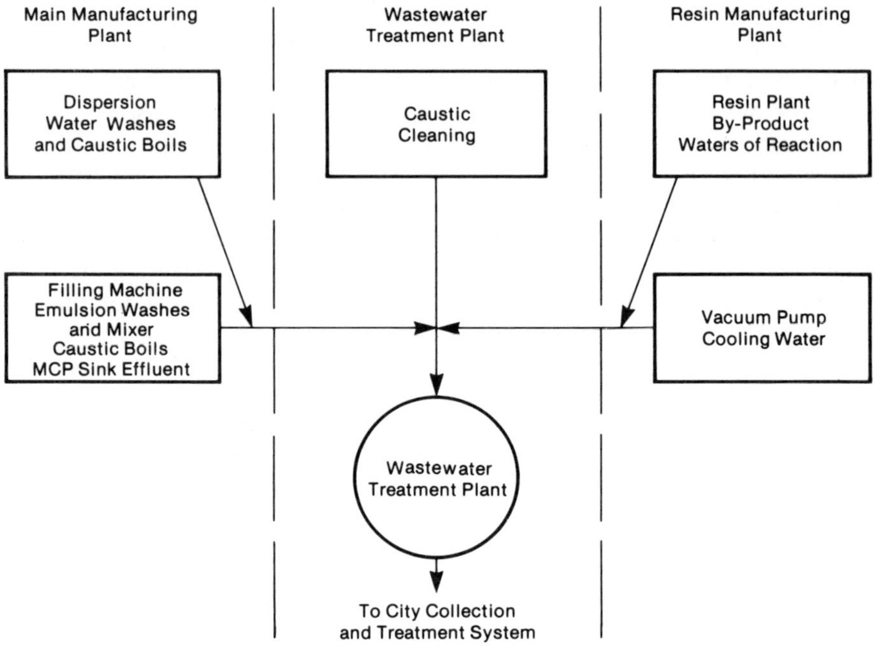

Figure 17-2. Pretreatment System

numerous changes in the parameters monitored, the structure of the charge formula, and the rates of surcharge. Perhaps most significant was the 1975 elimination of BOD from the charge formula. The BOD measure was replaced by the COD parameter. This was based on the sewer district's belief that COD was a better measure of waste loadings, due to the interference with the accuracy of the BOD test caused by some types of industrial discharges. A second significant change was the elimination, in 1977, of minimum strengths above which charges apply.

Strength-monitoring by the district is performed at least quarterly, with a composite sample taken each time. The composite sample consists of grab samples taken four times per day at approximately two-hour intervals between 9:00 A.M. and 5:00 P.M. All testing for constituents is done by district-owned facilities. Annual strength-related charges are based upon a simple average of observed concentrations from quarterly samples. Self-monitoring reports from certified laboratories can be used to supplement the samples.

The flow volume is based upon metered water-supply use. A determination can be made, based on a judgment by the superintendent of public works, of the amount of metered water that is lost in consumptive uses.

Alternatively, the industrial plants have the option of installing effluent flow meters to directly measure rather than estimate wastewater discharge. Strength charges are calculated from average concentrations and the estimated discharge volume.

Beginning in 1975, a permit system was instituted under a source-control ordinance. Permits are issued annually and place certain restrictions on the plant's use of the municipal wastewater collecting and treatment system. The ordinance limits permissible discharge concentrations of a variety of materials. Permit specifications include limits on the maximum daily flow, as well as maximum COD and TSS concentrations, and specification of a self-monitoring program of testing for COD, TSS, and a variety of restricted materials. The intent of the ordinance is to insure that the district is able to meet its own effluent-quality requirements and to encourage pretreatment by industry in compliance with EPA's point-source guidelines. Restricted constituents include toxic materials, heavy metals, and a number of compounds that are incompatible with the municipal biologial-treatment facilities.

This source-control program has posed severe constraints in several industries. Although not the sort of economic incentive under study in this research, the source-control restrictions have motivated consideration, and in some cases implementation, of significant wastewater-management responses by industry. In the process of trying to control restricted discharges, some permissible materials (for instance, COD, and TSS) are typically removed. When this has occurred, the resultant sewer-charge reductions are largely secondary to the primary goal of compliance with the source-control ordinance. Care is taken to discriminate among motivating factors in observed responses by the industrial plants under study.

During the period 1976 through 1977, much of the western United States experienced a significant drought. All of the plants in the city receive water supply from an affected utility. During much of 1977, the water utility had imposed a target of 25-percent cutback in water consumption. This target reduction was passed through to the industries as a voluntary water-conservation measure. In complying with this request, wastewater flows were significantly reduced in many cases. Although an associated sewer-charge reduction was achieved, it was not considered in all cases, or was a secondary factor to the primary goal of reduced water consumption. The specific plant responses taken will be discussed in the next section.

Response to Charges

On a day-to-day basis, the plant engineer of the facility has primary responsibility for environmental controls related to waste effluents. In addition, the plant manager has been active in the decision-making process. Decisions related to the management of the plant's residuals have been made almost exclusively at the plant level. There appears to be a general awareness that the charge program exists. The plant's effluent charges are currently about $4,500 annually. The cost effectiveness of providing various degrees of wastewater control as opposed to paying the charges has been considered and has resulted in some alterations in the plant's residuals-management practices, but most of the plant's efforts have been directed toward controlling constituents restricted under the city's source-control ordinance. The charge program has not induced changes in the types of commodities manufactured or the product mix.

Prior to 1961, the plant met a city requirement for the separation of storm and sanitary/industrial sewers. In addition to storm runoff, cooling water and effluent from the plant boilers were discharged to the storm sewer. The inclusion of nonrunoff effluent was a voluntary effort to reduce the volume upon which charges are based.

The plant's major response to the charge system occurred in 1961, with the installation of a wastewater pretreatment system and a calibrated effluent pump. The pretreatment scheme was designed primarily for TSS removal. Ferric sulphate was used as a coagulant, and aeration was used in the flocculation tank to make solids separation more effective. Some COD and BOD removal also occurred. Upon approval by the city, the effluent pump was used to directly measure waste volume, rather than estimating it based on metered water supply. Approximately 94 percent of the plant's water supply is not discharged to sewers.

Since the original plant was installed, a number of modifications have been made. Acid neutralization, recycling, and improved aeration were added in 1971. Presently, wastewater is collected in a 4,000-gallon tank. A 50-gallon-per-minute transfer pump sends the mixed waste through the system, where neutralization, flocculation, and aeration-standard methods separate the sludge and displace it into a holding tank. It is estimated that this system could treat 50 percent more wastewater than present production requirements demand. All wastewater, with the exception of domestic, is processed at the waste-treatment plant before entering the city sanitary sewer system. The coagulant used for solids removal has been changed a number of times over the years. Tests are presently being conducted with a number of new coagulants. Most of the upgrading of the original pretreatment system was the result of an effort to control wastewater characteristics other than COD and TSS.

Aside from the installation of a pretreatment system and effluent-measuring device, most other plant actions were not primarily in response to the sewer charges but were in reaction to an abatement order issued under the source-control ordinance. In August 1975, the superintendent of sanitation issued an order requiring the plant to take steps that would bring it into compliance with the city's source-control ordinance. The plant effluent was not complying with the requirement for toxicity, total chromium, and chlorinated organic compounds.

The plant's initial reactions to this order were:

1. To install a bioassay testing unit in the plant for monitoring of the toxicity problem.
2. To perform further tests to determine the source of chromium, which was not easily identified (testing of chemical flocculents for chromium removal was also being conducted).
3. To reduce the use of paint remover (dichloromethane) in the cleaning of portable tanks, which reduced the concentration of chlorinated hydrocarbons to one-tenth the level reported in the abatement order.

The plant has since identified the major source of the chromium problem: pigments in the solvent-type paints that enter the effluent stream in the caustic cleaning operations. A chemical (polymer) coagulant is being used in the pretreatment system for toxicity reduction and further chromium reduction. Presently, the plant is in compliance with the source-control ordinance most of the time.

The final response by the plant has been an active participation in a local manufacturers' association concerns group. The group has developed a number of documents—notably, some critical of certain provisions of the city's source-control ordinance. In addition, the group has provided carefully considered advice and review concerning engineering studies commissioned by the district. Most recently, the group has been engaged in an ongoing dispute with the sewer district concerning the validity of COD rather than BOD as the basis for sewer charges. This is unresolved as yet.

The plant is required to participate in the self-monitoring program under the source-control ordinance. The plant must test quarterly for COD, TSS, total chromium, lead, zinc, and toxicity (96-hour LC_{50}). Presently, the plant tests weekly for these constituents. It is anticipated that samples gathered will provide a broader database for the determination of charges.

The history of the plant's wastewater discharges to the municipal system is presented in figure 17-3. Yearly loadings for flow as well as COD and TSS concentrations are shown for the period 1973 through 1978 (figure 17-4). As mentioned earlier, production has remained fairly constant over the period. Overall, there has been no significant change in waste strength, as measured by COD and TSS, over the period. This behavior is expected, as the changes made to the plant's pretreatment system have not substantially altered its COD or TSS removal efficiency.

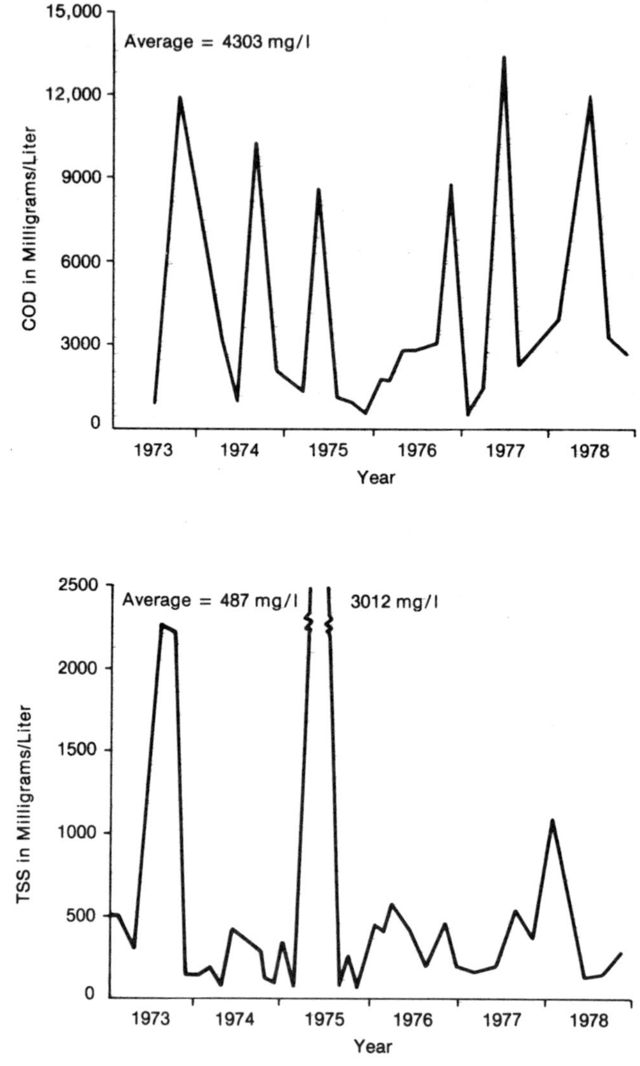

Figure 17-3. COD and TSS Histories—Paint Manufacturing

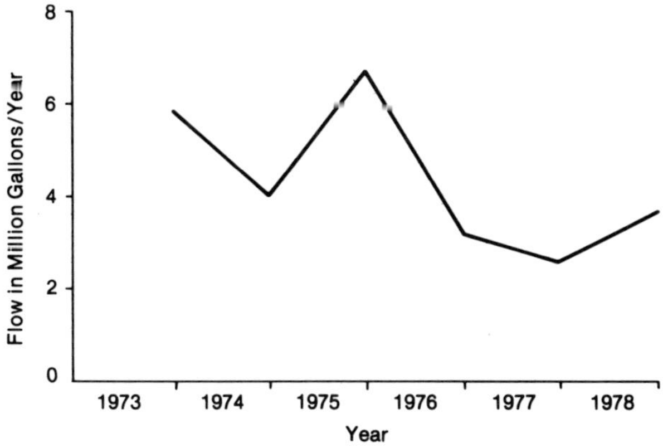

Figure 17-4. Flow History—Paint Manufacturing

Appendix: Other Literature in the Field

Ackerman, S.R. "Effluent Charges: A Critique." *Canadian Journal of Economics* 6 (4):512–528, November 1973.

Anderson, Norval E., and Sosewitz, Ben. "Chicago Industrial Waste Surcharge Ordinance." *Journal of the Water Pollution Control Federation* 43(8), August 1971.

Anderson, R.C., et al. "Environmental Charges: Economic, Technical, Legal, and Political Aspects." Washington, D.C.: National Academy of Engineering, Committee on Public Engineering Policy (draft report), July 1975.

Baumol, W.E., and Oates, W.J., "The Instruments for Environmental Policy." Presented at Conference on Economics of the Environment, November 10–11, 1972. National Bureau of Economic Research, Inc., New York.

Biniek, J.B. "Pollution Taxes, Effluent Charges, and Other Alternatives for Pollution Control." Issue Brief No. IB77009. Congressional Research Service, Library of Congress, Washington, D.C. February 8, 1977.

Bower, B.T., Ehler, C.N., and Kneese, A.V. "Incentives for Managing the Environment." *Environmental Science and Technology* 11(3), March 1977.

Brill, E.D., Jr., and Nakamura, M. "A Branch and Bound Method for Use in Planning Regional Wastewater Treatment Systems." *Water Resources Research* 14(1), February 1978.

Bubbis, M.S. "Industrial Waste Control in Metropolitan Winnipeg." *Journal of the Water Pollution Control Federation* 35(11), November 1963.

Buchanan, M., and Tullock, G. "Polluters' Profits and Political Response: Direct Controls Versus Taxes." *American Economic Review* 65(1): 139–145, March 1975.

Cabrera-Fernãndez. "El Desarollo Econбmico de Tijuana en Relaciбn al Suministro de Agua y a la Contaminacion Atmosférica, Marina Y Acuãtica." *Natural Resources Journal* 18(1), January 1978.

"Cities Treat Industrial Process Wastes." *Environmental Science and Technology* 5(10):1000–1002, October 1971.

Council on Law-Related Studies. "Effluent Charges on Air and Water Pollution: A Conference Report." Environmental Law Institute, Washington, D.C., 1973.

de Lucia, R.J. "An Evaluation of Marketable Effluent Permit Systems; Final Report." United States Environmental Protection Agency, Office of Research and Development, September 1974. (Distributed by National Technical Information Service, Springfield, Va., as PB-239 418.)

de Lucia, R., McBean, E., and Harrington, J. "A Water Quality Planning Model with Multiple Time, Pollutant, and Source Capabilities." *Water Resources Research* 14(1), February 1978.

Donahue, C. "Lawyers, Economists, and the Regulated Industries: Thoughts on Professional Roles Inspired by Some Recent Economic Literature." *Michigan Law Review* 70:195, November 1971.

Downing, Paul B. *The Economics of Urban Sewage Disposal.* Frederick A. Praeger, New York, 1969.

Eckstein, Otto. "The Allocation of Joint Costs," chapter IX of *Water-Resource Development—The Economics of Project Development.* Harvard University Press, Cambridge, 1958.

Elliott, Ralph D. "The Industrial Demand for Water and Waste Treatment in Selected United States Cities which are Levying Surcharges." Ph.D. thesis, Department of Economics, North Carolina State University, Raleigh, N.C., 1972.

Elliott, Ralph D., and Seagraves, James A. "The Effect of Sewer Surcharges on the Level of Industrial Wastes and the Uses of Water by Industry." Report No. 70, Water Resources Research Institute of the University of North Carolina, Raleigh, N.C., August 1972.

Ethridge, D.E. "An Economic Study of the Effect of Municipal Sewer Surcharges on Industrial Wastes." Report No. 41, University of North Carolina at Chapel Hill, Water Resources Research Institute, November 1970.

Freeman, A.M., III, and Haveman, R.H. "Residual Charges for Pollution Control: A Policy Evaluation." *Science* 177(4045):322–329, July 28, 1972.

Hanemann, Michael. "An Economic Analysis of Municipal Sewer Charges for Industrial Wastes" (draft). Department of Agricultural Economics, University of California at Berkeley, 1977.

Hoskins, W.L. "Let the Pricing System Provide the Incentive." *Industrial Water Engineering* 8(4):8–10, April 1971.

Hudson, James F. "No. 9: Design of Local User Charge Systems." *Papers in Support of the Resource Conservation Committee.* Council on Environmental Quality, 1979.

Hudson, J.F.; Bower, B.R.; and Brady, G. *Practical Experience with Implementation Incentives for Environmental Quality Management.* USR&E for CEQ, 1981.

Joint Committee of the American Society of Civil Engineers et al. "Fundamental Considerations in Rates and Rate Structures for Water and Sewage Works." *Ohio State Law Journal* 12(2), Spring 1951.

Kneese, A.V. (presiding). "Taxation and Management of Environmental Quality: First Concurrent Conference Session." National Tax Association, Columbus, Ohio, September 1971.

Kneese, A.V., and Bower, B.T. "Environmental Quality Analysis: Theory and Method in the Social Sciences." The Johns Hopkins Press, for Resources for the Future, Baltimore, 1972.

Kneese, A.V., and Schultze, C.L. "Pollution, Prices, and Public Policy." Brookings Institution, Washington, D.C., 1975.

Lane, L.L. "Taxing the Environment's Polluters." *People and Taxes* 4(1): 10–11, January 1976.

McLamb, David Wayne. "Industrial Demand for City-Supplied Water and Waste Treatment." Ph.D. thesis, Department of Economics and Business, North Carolina State University, Raleigh, N.C., 1978.

Marshall, D. "Environmental Protection and the Role of the Civil Money Penalty: Some Practical and Legal Considerations." *Environmental Affairs,* Spring 1975.

Maystre, Yves, and Geyer, John C. "Charges for Treating Industrial Wastewater in Municipal Plants." *Journal of the Water Pollution Control Federation* 42(7), July 1970.

Meta Systems, Inc. "Effluent Charges: Is the Price Right?" United States Environmental Protection Agency, Washington, D.C., 1973

Michel, Robert L. "Costs and Manpower for Municipal Wastewater Treatment Plant Operation and Maintenance, 1965–68." *Journal of the Water Pollution Control Federation* 42(11), November 1970.

Milliman, Jerome W. "Beneficiary Charges: Toward a Unified Theory." In *Public Prices for Public Products,* Selma J. Mushkin (ed.). The Urban Institute, Washington, D.C., 1972.

Nagel, Stuart S. "Incentives for Compliance with Environmental Law." *American Behavioral Scientist* 17, May-June 1974.

Oates, W.E., and Strassman, D.I. "The Use of Effluent Fees to Regulate Public Sector Sources of Pollution: An Application of the Niskanen Model." *Journal of Environmental and Economic Management* 5(3), September 1978.

Ollife, Joseph J. "Sewer Service Charges and Surcharges." *Journal of the Water Pollution Control Federation* 35(5), May 1968.

Organization for Economic Cooperation and Development. "Paying to Pollute." *Environment* 18(5):16–20, June 1976.

Organization for Economic Cooperation and Development. "The Polluter Pays Principle: Definition, Analysis, Implementation." Paris, 1975.

Public Works Engineers. "Industrial Waste Disposal Charges in Cities Over 5,000 Population." American Public Works Association, Special Report No. 18-S, January 1955.

Quirk, Thomas P. "Economic Decision Marking for Participation in Joint Industrial-Municipal Treatment Systems." Presented at the National Symposium on Costs of Water Pollution Control, sponsored by the

Research Triangle Universities and the U.S. Environmental Protection Agency, Hilton Inn, Raleigh, N.C., 1972.

Reitze, A.W., and Reitze, G. "Tax Incentives Don't Stop Pollution." *American Bar Association Journal,* February 1971.

Renshaw, E.F. "Pollution Taxes: Political Realities May Require Rebates." *Tax Notes* 4(42):3–4, 10–14, October 18, 1976.

Renshaw, E.F. "Should the Federal Government Subsidize Industrial Pollution Control Investments?" *Journal of Environmental Economics and Management* 1(1):84–88, May 1974.

Ricketts, M., and Webb, M.G. "Pricing and Standards in the Control of Pollution." *Scandinavian Journal of Economics* 80(1), 1978.

Roderick, Ralph E. "Rate Structures for Industry." *Journal of the Federation of Water Pollution Control* 34(4), April 1962.

Rose-Ackerman, S. "Market Models for Pollution Control: Their Strengths and Weaknesses." Paper prepared for the National Commission on Water Quality, Yale University Department of Economics, New Haven, September 1975 (unpublished).

Ruff, L. E. "The Economic Common Sense of Pollution." *The Public Interest* 19:69–85, Spring 1970.

Russell, C.S. "What Can We Get from Effluent Charges?" *Policy Analysis* 5(2), Spring 1979.

Seagraves, J.A. "Puzzled About Surcharges? Here's How Durham Operates." *Waste Water Engineering* 9(3), March 1972.

Seagraves, J.A. "Sewer Surcharges and Their Effect on Water Use." Paper presented at the annual meeting of the North Carolina Water Pollution Control Association, Durham, November 1971.

Shaw, Ray E., Jr. "Experience with Waste Ordinance and Surcharges at Greensboro, N.C." *Journal of the Water Pollution Control Federation* 42(1), January 1970.

Smith, Robert. "Cost of Conventional and Advanced Treatment of Wastewater." U.S. Department of the Interior, FWPCA, Cincinnati, 1968.

Symons, George E., and Crane, Frederick W. "Special Sewer Service Charges for Industrial Wastes." *Water Works and Sewage* 91(3), March 1944.

Symons, James M. "Rate Formulas for Industrial Wastes." *Water and Sewage Works* 101(12), December 1954.

Tihansky, D. "A Survey of Empirical Benefit Studies." In *Cost Benefit Analysis and Water Pollution Policy,* H.M. Peskin and E.P. Seskin (eds.). The Urban Institute, Washington, D.C., 1975.

Tunderman, D.W. "Economic Enforcement Tools for Pollution Control: The Connecticut Plan." Presented at the Environmental Study Conference Briefing, Tax and Fee Approaches to Pollution Control, Washington, D.C., May 26, 1976.

Tullock, G. "Excess Benefit." *Water Resources Research* 3(2):643–644, Second Quarter 1967.

Tybout, R.A. "Pricing of Pollution." *The Bell Journal of Economics and Management Science* 3(1):252–66, Spring 1972.

U.S. Department of the Interior, Environmental Protection Agency. "Proposed Rules—User Charges and Industrial Cost Recovery." *Federal Register* 38(98), May 22, 1973.

U.S. Department of the Interior, Environmental Protection Agency, Office of Water Programs. "Federal Guidelines: Equitable Recovery of Industrial Waste Treatment Costs in Municipal Systems." October 1971.

U.S. Department of the Interior, Federal Water Pollution Control Administration. "The Cost of Clean Water and its Economic Impact." *Sewerage Charges* 3, January 10, 1968.

U.S. House of Representatives, Committee on Government Operations, "Views of the Governors on Tax Incentives and Effluent Charges." House Report 1330, Union Calendar No. 567, March 15, 1966, 89th Congress, Second Session. United States Government Printing Office, Washington, D.C., 1966.

Walker, M., and Storey, D.J. "The 'Standards and Price' Approach to Pollution Control: Problems of Iteration." *Scandinavian Journal of Economics* 79(1), 1977.

Walter, C. Richard. "Determination of Charges and Surcharges for Waste Treatment." *Journal of the Water Pollution Control Federation* 35(9), September 1963.

Wenner, L.M. "Pollution Control: Implementation Alternatives." *Policy Analysis,* Winter 1978.

Wright, Robert. "Sewage Service Charges." Agricultural and Mechanical College of Texas, Bulletin No. 98, 1947.

Yance, J.V. "Analysis of Differences in Municipal Treatment Levels." *Water Resources Research,* December 1978.

Index

Index

About the Authors

James F. Hudson received the Ph.D. in civil-engineering systems in 1975 from Massachusetts Institute of Technology, where he concentrated on economic issues relating to municipal-service demand and service delivery. At Urban Systems Research and Engineering, Inc., Dr. Hudson's primary interests have been land use and environmental-management issues.

Elizabeth E. Lake, a senior economist, has directed a number of economic and environmental analyses since joining Urban Systems Research and Engineering, Inc. She has extensive experience in economic evaluations and evaluations of the impact of federal policies and procedures. Prior to her present position, Ms. Lake worked as an economist for the Systems Research Group of Toronto and the Royal Commission on the Economic State of Prospects of Newfoundland. Ms. Lake did her post-graduate work in economics at the University of Toronto.

Donald S. Grossman, a systems-analysis consultant, has worked extensively in civil-engineering systems, policy-analysis techniques and applications, and the problems of investment and operation by municipal water and wastewater utilities. Dr. Grossman has served as manager of the Systems Management and Control Division of Resource Analysis, a CDM company, and was affiliated with the faculty of Massachusetts Institute of Technology. Dr. Grossman received the Ph.D. in civil-engineering systems in 1977.